Curing Disease from the Ground Up

Timothy M. Block

Curing Disease from the Ground Up

How to Operate a Biotechnology Business Incubator

Timothy M. Block
Baruch S. Blumberg Institute
Pennsylvania Biotechnology Center
Doylestown, PA, USA

Department of Pathology
Yale University School of Medicine
New Haven, CT, USA

ISBN 978-3-031-56150-4   ISBN 978-3-031-56148-1   (eBook)
https://doi.org/10.1007/978-3-031-56148-1

© The Editor(s) (if applicable) and The Author(s), under exclusive license to Springer Nature Switzerland AG 2024

This work is subject to copyright. All rights are solely and exclusively licensed by the Publisher, whether the whole or part of the material is concerned, specifically the rights of translation, reprinting, reuse of illustrations, recitation, broadcasting, reproduction on microfilms or in any other physical way, and transmission or information storage and retrieval, electronic adaptation, computer software, or by similar or dissimilar methodology now known or hereafter developed.

The use of general descriptive names, registered names, trademarks, service marks, etc. in this publication does not imply, even in the absence of a specific statement, that such names are exempt from the relevant protective laws and regulations and therefore free for general use.

The publisher, the authors and the editors are safe to assume that the advice and information in this book are believed to be true and accurate at the date of publication. Neither the publisher nor the authors or the editors give a warranty, expressed or implied, with respect to the material contained herein or for any errors or omissions that may have been made. The publisher remains neutral with regard to jurisdictional claims in published maps and institutional affiliations.

This Springer imprint is published by the registered company Springer Nature Switzerland AG
The registered company address is: Gewerbestrasse 11, 6330 Cham, Switzerland

If disposing of this product, please recycle the paper.

Dedicated to my wife, Joan and our two children, Peter and Suzanne, and their spouses Christina and Patrick. In memory of my parents, Carl and Joyce.

Preface

If you had little money, but a compelling personal drive to cure a disease that was of little interest to big pharma, there are few options available. What can you do? (i) Find the cure yourself? (ii) Move public policy such that the federal government would fund cure research? (iii) Start your own research foundation and technology center? This book asks that question and provides an example, as unlikely as it sounds, where all three of those options were tried. In providing these explanations, the book describes the regional and national biotechnology business climate that existed during the time the technology center and research programs described in this book occurred. Therefore, one important part of this book is a brief biotechnology rationale for a life sciences incubator as well as, literally, a "how to" manual about starting and operating a biotechnology business incubator.

In short, the book has two goals. It is a personal story about how learning someone close to me had a chronic disease caused my friends and me to start a nonprofit foundation (Hepatitis B Foundation) *and* a research institute (Baruch S. Blumberg Institute) dedicated to finding a cure and helping people affected by hepatitis B worldwide. However, the way that we decided to get these nonprofits funded is its own story. We planned to build a commercially oriented life sciences business incubator called the Pennsylvania Biotechnology Center (PABC). Hence, this is a biotechnology story. Our incubator's success has drawn attention from business professionals, elected officials, and members of the general public. They have asked about the reasons for its success and its operational details. Therefore, the book is divided into two parts, which should be of interest to the overlapping readerships.

The first part tells the story about why and how I started the PABC. This ties all three organizations together, showing the logic for bundling them, since at first glance they probably seem quite different. However, there is a rationale and quite a bit of complementarity. It was a way to fund the small nonprofit Hepatitis B Foundation (HBF) that we started in 1991 and its nonprofit research organization, the Baruch S. Blumberg Institute (Blumberg Institute), created in 2003. They were noble enterprises but did not attract significant funding. Our solution was to establish the PABC in 2006.

The path to the creation and operation of these organizations is also a journey through personal upsets, disappointments, the engagement of elected government representatives, big pharma VIPs, and, importantly, scientist celebrities, including several Nobel laureates. Many of the individuals involved are well-recognized figures, but many much less so, and they all get call outs because identifying them provides credibility for the story. Our success was also *favorably* affected by both the biotechnology revolution credited with the discovery of life sustaining and saving medicines and, surprisingly, the horrible event of 9/11 in which the entire country was traumatized by terrorist attacks.

Taken together, this stretch of time from the 1980s through 2000s in the USA could be considered the "Golden (if not Gilded) Age" of the biotechnology and molecular genetics revolution. That is, the foundational discovery of the structure and nature of DNA replication was reported by Drs. Watson and Crick in 1953. Molecular cloning was introduced in the 1970s. And by the 1980s and 1990s, the practical use of molecular biology for biotechnology products became apparent and prevalent, with many academic departments of biology and biochemistry rebranding themselves as departments of Molecular Biology. That was a wave we were riding, even though we may not have known it.

This all makes the first part of the book a somewhat personal story, beginning with details about the reasons for starting the PABC. Briefly, the basic idea was to create an environment of research companies that would focus on hepatitis B and liver diseases. My wife, who is Asian American, had been diagnosed with chronic hepatitis B, which can cause liver cancer and is prevalent in Asia. As I was a beginning member of a university faculty, we had little money ourselves to do anything about her illness. The book explains why this was our solution to that challenge. The first part of the book is the *"why we did it."*

A friend in paraphrasing a famous author said, every good story needs love, adversity, a villain, and a hero, so this book does have a bit of each. The personal story about why we started the PABC, the difficulties we faced, and we can even add a few heroes.

However, the PABC has a story and constituency of its own. It is now one of the most successful life sciences incubators in the USA and has had an impact in the biotechnology business sector, which is quite distinct from the HBF and distant from our original purpose. Chapter 2 discusses how the PABC got there. Startup companies from the PABC have led to pioneering therapies for chronic hepatitis B in clinical trials (our original intention), but also medicines and medical devices in use for other disorders, such as gastrointestinal disease, cancer, and Alzheimer's disease. This is in addition to creating thousands of jobs and several billion dollars of investor value.

Another surprise: we are located in Doylestown, Pennsylvania, which is distant from the areas of the country usually associated with high tech corridors, but the PABC is still consistently ranked in the top ten US incubators by rating agencies. Therefore, there has been considerable interest in how that came to be and in the actual operational mechanics of our place. Chapter 2 of the book addresses the biotechnology and entrepreneurial climate that made it possible for the PABC to thrive and how we leveraged this innovation revolution. It's the *"how we did it."*

Chapter 3 of the book is a detailed description of the PABC and its operations. This is intended as an example of a business incubator and literally provides a Biotechnology Center Operations Manual. When we created the PABC, and even today, there were few such sources of information despite the vast resources of the internet. There are now more than 1,200 business incubators in the USA, perhaps 6000 worldwide, yet there are few, if any, books about them or guidance about how they can function. The hope is that this book can fill a gap. It's a *"how to do it."*

Taken together, the book contains facts and details about how to run a Biotechnology Incubator, as well as personal stories about why to start and run a biotechnology center. Gathering these facts and remembering these stories involved speaking with and help from several people. Much of the book depends upon my recollection of events and must, therefore, be considered with that qualification. Perhaps that is why I appear to be the "hero" of many of these stories. That said, where statistics about biotechnology incubators are provided, I have tried to be more rigorous, providing citations to support whatever facts I am asserting. However, the scholarly literature and validated research about biotechnology incubators are rather sparse and often depend upon internet-based sources. Fortunately, credible information can be found at sites such as the International Business Innovation Association (InBIA) and the Ewing Marion Kauffman Foundation.

The internet and the scholarly literature were helpful in writing this book, but so were many individuals. I am grateful to my wife, Joan, for her helpful reading of the manuscript. I am not sure if this book will make it to her book club, but she did read the manuscript critically. She has been my partner in our family, career, and in establishing each of the three organizations described in this book. Lou Kassa, my successor as CEO of the three nonprofits, has helped craft and implement the unique business model that the PABC takes. He provided drafts of the PABC business outline described. Konrad Kroszner, PABC Director of Systems and Engineering, and Lorretta Molle, PABC Director of Business Administration, provided helpful information about the Center's infrastructure and policies, which were used in the book. Chari Cohen, DrPH, MPH, and current President of HBF provided inspiration and encouragement for this book as well. I must recognize the very helpful comments of Matteo Prayer, from Manuscripts Press, who was insightful and greatly improved the text readability. Finally, I want to recognize Judith Marchand, Blumberg's Director of Corporate Governance and Legal Affairs, with whom I have worked closely over the past decade, and express gratitude for her patience and careful editing.

Doylestown, PA, USA Timothy M. Block
New Haven, CT, USA

Contents

1	"Starting It: Why I Started a Life Sciences Incubator"	1
	1.1 Why Should We Build?	1
	1.2 Entrepreneurial Philanthropy	7
	1.3 A Leafy Liberal Eden	12
	1.4 9/11 and Creation of the Pennsylvania Biotechnology Center	18
	1.5 The Pennsylvania Biotechnology Center is Located in the Best and Worst Place	27
	1.6 The Biotechnology Industrial Revolution Happened Despite Scientist Self-Consciousness	28
	1.7 Celebrity Science: Baruch S. Blumberg and W. Thomas London	31
	1.8 The Leap: Leaving the University	33
	1.9 The Golden Age of HBV Drug Discovery	36
	1.10 Initial Public Offerings on Research Way Boulevard	42
	1.11 A Good Time to Be a Virologist But Not So Good for the World: COVID-19	44
	1.12 Growing During the COVID-19 Pandemic	47
	References	50
2	Biotechnology and Entrepreneurship Need Business Incubators	51
	2.1 Entrepreneurism and Innovation	51
	2.2 Small Business as a Vehicle for Innovation	55
	2.3 Startup Company Anatomy	61
	2.4 The Business Incubator Concept	67
	2.5 Categories of Business Incubators	71
	2.6 Advantages to the Entrepreneur and Community of a Life Sciences Business Incubator/Accelerator	72
	2.7 Defining Success for Life Sciences Incubator	73
	2.8 Why Do Startups Fail?	78
	References	80

3	**Pennsylvania Biotechnology Center: A How-to Manual**............	83
	3.1 Description of the Campus and Its Physical Layout	84
	3.2 Power, Water, Information Technology (IT), and Infrastructure Needs of the Incubator....................	85
	3.3 Management of the PABC.................................	87
	3.4 The Ins and Outs of Being a Tenant Member	91
	3.4.1 Choosing Tenants, Advertising	91
	3.4.2 Moving Tenants In and Out.........................	93
	3.5 Tangible and Intangible Support to Startups at the PABC........	94
	3.6 To Whom and For What Is the PABC Accountable?............	99
	3.7 Taking Equity in the Startup Companies at the PABC	101
	3.8 Graduate Space and Replicating the Model...................	102
	3.9 Taking the PABC Model On the Road: Replicating the Model....	103
	References...	105
Afterword...		107
Appendices...		109
Index...		111

List of Figures

Fig. 1.1	Image of the cover of Time magazine from 1977, 1980	29
Fig. 1.2	NIH Funding for HBV (in millions of $)	37
Fig. 1.3	Biotech success stories from the Pennsylvania Biotechnology Center	43
Fig. 2.1	Number of U.S. FDA Drug approvals per year and country of origin	56
Fig. 2.2	Deals as venture investments (U.S.) 2013–2023	61
Fig. 2.3	Number of Business Incubators and Accelerators. Total number of incubators (**A**) and the subset of business incubator self designated as "accelerators" (**B**) is shown. Numbers of business incubators in the U.S	70
Fig. 2.4	Startup activity by State (**a**) relative amount of start-up activity in larger states (**a**) and smaller states (**b**)	70
Fig. 2.5	Categories of business incubators in the U.S	72
Fig. 3.1	Panoramic of the campus (**a**). Floor plan (**b**) and profile of the Baruch S. Blumberg Institute and accelerator wing (**c**)	85
Fig. 3.2	Organizational chart of BSBI, HBF and PABC	88

List of Tables

Table 1.1	Demography of the home of the Pennsylvania Biotechnology Center	28
Table 2.1	2019 Global Entrepreneurial Index (top 20 and bottom 20 countries shown)	54
Table 2.2	Selected top selling/medically important medicines/advances originating in small companies	57
Table 2.3	2022 Notable deals between big pharma and smaller companies	58
Table 2.4	Smaller strategic deals in 2022	59
Table 2.5	Notable deals between big pharma and smaller companies (2020/21)	60
Table 2.6	Biotech deals in the first quarter of 2024	62
Table 2.7	Biotech deals of 2023 (last quarter 2023 shaded)	63
Table 2.8	Private equity cash reserves (2014–2023)	65
Table 2.9	The 15 countries with the most start-ups	65
Table 2.10	Categories of startup companies in the U.S	66
Table 2.11	How have the demographics of entrepreneurs who start new companies changed over time?	67
Table 3.1	Occupant/tenant profile at the PABC	85
Table 3.2	Space distribution of the PABC	86
Table 3.3	Staff managing the PABC, BSBI, and HBF	90
Table 3.4	Recommended "Punch List" of actions to be taken to launch each "Newco"/Portfolio Company (managed by the Portfolio Review Committee)	97

About the Author

Timothy M. Block, PhD, Dr. Timothy Block is Visiting Professor, Yale University School of Medicine and Chairman, Co-founder and President Emeritus of the Hepatitis B Foundation (est. 1991); Founder and President Emeritus of the Baruch S. Blumberg Institute (est. 2003); and Founder and President Emeritus and (ret.) CEO of the Pennsylvania Biotechnology Center (est. 2006).

He is scientific co-founder of several life sciences companies, served on numerous company Boards, ranging from those that are startups to publicly traded, and helped in their raise of hundreds of millions of dollars. He is co-inventor on 20 issued patents and 23 applications, has co-authored more than 270 scholarly papers, received more than $50 million in extramural research during his career, has advised and served on numerous federal review panels (NIH, FDA, DoD), and continues to be an adjunct professor at the University of Pennsylvania School of Medicine. He has received numerous awards and honors that include being inducted as Fellow into the American Association for the Advancement of Science (AAAS) and Fellow, U.S. National Academy of Inventors; special citations from the U.S. Congress and PA legislature; "Best Life Science CEO" from the *Philadelphia Business Journal* and PharmaVoice 100; "Researcher of the Year" from the American Liver Foundation and named Fellow of the American Association for the Study of Liver Diseases (AASLD). In 2020, he and his wife received the inaugural AASLD Distinguished Public Service Award on behalf of the Hepatitis B Foundation.

Tim and Joan Block accepting the AASLD national public service award for the HBF (2020).

Abbreviations

BSBI	Baruch S. Blumberg Institute
CDC	U.S. Centers for Disease Control and Prevention
CFTR	Cystic Fibrosis Transmembrane-conductance Receptor
Del Val	Delaware Valley College, currently Delaware Valley University
Drexel	Drexel University
FDA	U.S. Food and Drug Administration
GEDI	Global Entrepreneur Development Institute
HAV	Hepatitis A Virus
HBV	Hepatitis B Virus
HCV	Hepatitis C Virus
HBF	Hepatitis B Foundation
KIZ	Keystone Innovation Zone
NASA	National Aeronautics and Space Administration
NB-DNJ	N-butyl-deoxy-nojirimycin
NBIA	National Business Incubator Association
Newcos	New companies
InBIA	International Business Incubator Association
NIH	U.S. National Institutes of Health
PABC	Pennsylvania Biotechnology Center
SAB	Scientific Advisory Board
SAM	Systems for Awards Management
SBA	U.S. Small Business Administration
SBIR	Small Business Innovation Research
STTR	Small Business Technology Transfer
TJU	Thomas Jefferson University College of Medicine
U. Del	University of Delaware
U.S.	United States

Chapter 1
"Starting It: Why I Started a Life Sciences Incubator"

"It is just a pile of rocks until someone figures out how to build the cathedral ..." paraphrasing Antoine de Saint-Exupéry in his book *The Little Prince*.

1.1 Why Should We Build?

I don't think it's an exaggeration to write that modern civilization exists and is expressed through its institutions. That may be a bit unfashionable or pretentious to say, but to the contrary, there may even be a humility in recognizing that public, private, and social institutions are manifestations of and actually are the vehicles by which societies and civilizations function. I am thinking of libraries, hospitals, universities, and, yes, government agencies. Specific examples are the U.S. Post Office and the National Institutes of Health. Religious institutions ranging from country churches to the Vatican are included. A society's technology, materials, and social interactions are organized around and through its institutions. The products and amenities of civilization exist within and because of their structures. These institutions do things that individuals, on their own, would be unable to do. Construction of glorious buildings and solving complex, tremendous problems are just a few examples.

Institutionalism is an entire branch of social sciences and sociology (Kostova et al. 2008; Peters 2019). I consider myself an institutionalist with a great respect for institutional structures. This was not always the case. Initially, as a young faculty member beginning my career at Thomas Jefferson University College of Medicine (TJU), I was cynical, even suspicious, about institutions and their bloated bureaucracies and clumsiness. I valued what I could do myself way above the enterprise in which I worked. Today, 40 years later, I am writing this from my office at Yale University, where I am spending a couple of years as a visiting professor. I am struck by the importance, capability, grandeur, and expectations we have of a great institution such as Yale. In addition, frankly, how limited the contributions would be from an individual working alone without the benefit of cohesive, resource-providing institutions.

Using a biology analogy, a single cell can carry out remarkable functions, and there are many things that can be accomplished by an individual alone. However, there are many particularly complex activities that require divergent but coordinated functions, such as what we observe in multicell organisms. It may be a bit off-putting to think of our civilization as a multicell organism, with the image of a sponge first coming to mind, but the analogy holds.

What a society has to offer, what it can create, how it consumes, and how much of what it consumes are done through its institutions. In a small way, the Pennsylvania Biotechnology Center (PABC), which is the focus of this book, is an example of mini-institutions that I have come to recognize and appreciate as playing a significant economic and social role in our region.

Comparing a small business incubator with a centuries-old academic institution may be slightly presumptuous, but the role of small institutions in society should not be undervalued. Much of society functions through small businesses, schools, and microstructures. The business incubator, I have come to realize, plays a valuable role.

In 1990, I wanted to create an institution: the Hepatitis B Foundation (HBF). The very ambitious and aspirational goal of the HBF was to find a cure for diseases caused by the hepatitis B virus (HBV). It was need, more than ambition, that motivated me. My wife, Joan, was born in South Korea and infected at birth with HBV, but she wasn't diagnosed with chronic hepatitis B until adulthood. People with chronic hepatitis B have a very significant risk of dying prematurely from liver cirrhosis or liver cancer. For example, the risk of a person with chronic hepatitis B dying from liver cancer is greater than the risk of a heavy smoker dying from lung cancer (Iloeje et al. 2007).

Although there was a good vaccine as early as 1986 to prevent people from developing chronic hepatitis B, there were no medicines to cure people once they had the disease. We learned about Joan's chronic hepatitis B from an employee physical when she was applying to work as a nurse in a local hospital. Up until that time, she did not realize her situation. Even today, a surprise diagnosis is very typical. Most people with chronic hepatitis B do not know they are infected until well into their 40s, 50s and 60s when the disease rudely declares itself. Until then, it is usually a silent infection. Joan was 30 years old, and we were just getting started with our married life. This isn't to pull at anybody's heartstrings, but learning of Joan's condition was a shock for us. Making things even more frightening was that Joan had been caring for a young woman in her 30s dying from hepatitis B-related liver cancer because there were no drugs for either condition.

Although my career was just at its beginning, I was at TJU, a medical college in Philadelphia, where I thought I was in a position to do something. I wanted to do something big. I turned to a couple I had come to know for their creativity and humanitarianism. Paul Witte was a successful design engineer, credited with designing everything from popular tennis rackets to orthopedic implants. He lived in the bohemian artist community of New Hope, Pennsylvania, in Bucks County. To me,

Paul seemed to be a giant, not just in physical stature (he was very tall) but also in intelligence and wit. He had prominence, culture, and a well-off entrepreneurial lifestyle filled with intellectualism, art, and entertainment. His wife, Jan, was also highly intelligent, caring, and a well-known supporter of many struggling artists, writers, and other stray young people in the area. I was enamored. I told them I wanted to do something big: find a cure for chronic hepatitis B.

Paul's 5-year-old son from a first marriage had passed away from another disease decades prior, so both he and Jan were extremely sympathetic to my story. We decided, initially, to try to get big pharma to work on a cure for HBV. Why not? In the 1990s, there were an estimated 300 million people in the world who were chronically infected. It seemed to be an enormous market opportunity, we reasoned. Therefore, before starting a new nonprofit foundation, we thought we would try to entice pharma and the federal government to invest in HBV research since they would obviously have the means we did not. To add some star power to our outreach, I had also approached Baruch S. Blumberg, M.D., Ph.D., the Nobel Prize-winning scientist who had discovered HBV and developed its first vaccine and was working at Fox Chase Cancer Center in Philadelphia. To my surprise, he had moved to England, where he was serving as Master (i.e., Dean) of Balliol College at the University of Oxford. Fortunately, he regularly returned to the U.S. I asked if he would help the cause by either starting a foundation for a cure or by helping us persuade big pharma to take an interest. He was all in without any reservations or demands. With Dr. Blumberg's stature, we were able to get meetings with executives at several major pharmaceuticals. We made the case that they should be working on developing drugs to treat HBV. Given that 300 million people worldwide suffered from chronic hepatitis B, I thought the argument was rather obvious.

The executives at each of the companies (I think we visited three) were polite. They heard us out. In addition, they all had pretty much the same response: HBV was a major problem, but it was mostly prevalent in countries outside North America and Europe. Their companies set strategic priorities based on a combination of factors that included their resources and, of course, the likely market. In 1990, China, despite having more than 100 million people living with chronic hepatitis B, was not considered an attractive market. Therefore, no. They were not going to develop HBV medicines any time soon.

We went to the American Liver Foundation and proposed raising funds that would be dedicated to an HBV cure. They declined our offer to create a specific fund for this. There was nowhere else to go.

I had come to know our U.S. Congressman from Bucks County, PA, Jim Greenwood, before he was elected to office. He was initially a social worker and environmental activist. His first elected office was in the Pennsylvania state legislature, where he helped us with HBV vaccine legislation. Now, as a U.S. Congressman, he arranged for us to meet with leadership at the U.S. National Institutes of Health (NIH) and appeal to them to pay more attention to HBV. At the meeting, there were approximately four NIH deputy and division directors, representatives from the U.S. Surgeon General's office and the Centers for Disease Control and Prevention (CDC), the Congressman, and the Wittes and me. Congressman Greenwood, at one point, asked the federal leaders, "If I could reach into my wallet and give you whatever amount of money you need, what would you want to do for HBV?" One of the NIH leaders answered almost immediately. His response was startling. He said something to the effect, "I would NOT spend it on looking for new drugs for HBV. We have a good vaccine. If we vaccinate everybody there will be a 'generational wall' of protection, and in 30 years there will be no more HBV." Congressman Greenwood, perhaps seeing the look on my face, asked, "What happens to the people who have HBV now? Won't they die from liver cancer?" There was an awkward silence, but to my surprise, the individual from the NIH didn't back down. He suggested that it was essentially more efficient and more practical from a public health perspective to use the tools we have, such as the vaccine, to prevent all future infections. In 30 years, he said, "The problem would be solved with an immunized generational wall." In other words, this would be about the time it would take for everyone living with HBV in 1990 to die from its complications, such as cirrhosis and/or liver cancer.

It sounded heartless, especially with us in the room. However, there was, at the time, a prevailing public health logic in the position. I think Congressman Greenwood could see my eyes welling up. If he didn't appreciate our challenge before this meeting, he could see it now. He became a lifelong ally of the cause. This doctor, by the way, I have come to know as a remarkable, dedicated public health professional, and I know he would cringe if he remembered this story. Although he was clearly tone-deaf and undiplomatic at that meeting, his attitude was not unusual 30 years ago.

The "Generational Wall" strategy of *waiting for everyone with HBV to die and be replaced with a generation of immunized people* was unsettling. That approach would have condemned everyone in 1990 with HBV to live or die with their risk of serious liver disease, asking them to understand the impracticality or needlessness of a cure. That, of course, included my wife, Joan. And so many others we knew. Obviously, this was not a solution. The big step we decided to take after that meeting was to start a nonprofit research foundation.

To me, those were the dark ages of hepatitis B. Dr. Blumberg, with Harvey Alter, M.D., and Sam Visnich, Ph.D., first reported the discovery of HBV, initially calling it the "Australia antigen." It was supported by circumstantial, albeit powerful, evidence from Drs. Blumberg and Alter, and W. Thomas London, M.D., Bruce Smith, M.D., and others. The Australia antigen was determined to be the cause of chronic hepatitis B and ultimately the major etiology of liver cancer. Nearly one million people die each year worldwide from HBV-related liver cancer.

The path to the discovery of HBV and its association with liver disease was not straightforward. It remains one of the great stories in medicine. Dr. Blumberg was a medical anthropologist who traveled and collected blood from people from all over the world. In his research lab he identified a protein antigen that was originally thought to be associated with leukemia in the serum of people in Australia and Borneo. Hence, his group called the protein "Australia antigen." A few years later, they and others noticed that this antigen also turned up in the blood of people who developed "hepatitis," which means liver inflammation.

1.1 Why Should We Build?

The definitive event for identifying HBV is a well-known story in virology. Barbara Werner, a research assistant in Dr. Blumberg's lab, accidentally stuck herself with a needle containing Australia antigen. Her pre-needlestick blood test was negative for the antigen (they had archived her blood at the time she began working with Dr. Blumberg). But after the needlestick, she became symptomatic for hepatitis, and her blood test became "positive" for Australia antigen. Fortunately, as with most healthy adults who become infected with HBV, she completely recovered after a few months, becoming negative and gaining a protective antibody to Australia Antigen. This provided very convincing evidence that Australia antigen was the cause of what was then called "serum hepatitis." Increasing evidence that Australia antigen was the cause of serum hepatitis came later from studies of adults in prisons, institutional care centers, and children in orphanages. Institutionalized people developed serum hepatitis at very high frequencies, and the correlation between Australia antigen in the blood and the development of hepatitis (i.e., liver inflammation) could be made. Eventually, Australia antigen was renamed the hepatitis B virus.

The reason HBV was initially known as serum hepatitis was because of its transmission through blood. It was distinguished from the other types of viral hepatitis called "infectious hepatitis" (caused by the hepatitis A virus or HAV), which is transmitted from contaminated food and water. If that was not confusing enough, there was another increasingly common viral hepatitis, initially called "non-A non-B hepatitis" (now known as hepatitis C). This is because it was transmitted primarily through blood transfusions, yet tested negative for serum hepatitis (HBV) and infectious hepatitis (HAV). The actual hepatitis C virus (HCV) would not be identified until years later by Dr. Harvey Alter (yes, the same person who helped identify HBV) and Drs. Daniel Bradley, Qui-Lim Choo, and Michael Houghton. Drs. Alter and Houghton would go on to receive the 2020 Nobel Prize in Medicine for their role in this work, along with Charlie Rice, Ph.D.

Dr. Alter had hypothesized that there must still be a hepatitis virus out there because, even though HBV was screened out from the blood supply, transfusion-associated hepatitis continued to be a significant problem. He produced a panel of blood samples from people with and without non-A non-B hepatitis. Drs. Bradley, Choo, and Houghton were then able to prove that the viral proteins they molecularly cloned from the blood of chimps infected with non-A non-B hepatitis were the same. They could now call it the hepatitis C virus. Dr. Rice and others, such as Ralf Bartenschlager, Ph.D., then produced critical systems to study HCV and generate new drugs.

These systems were used by scientists such as Michael Sofia, Ph.D., at the startup company Pharmasset and Min Gao, Ph.D., at Bristol Myers Squibb to discover and produce curative drugs (sofosbuvir and daclatasvir, respectively) for HCV. This was considered one of the great triumphs of the partnership between clinical medicine and molecular biology and has become a historic model. Consider in 2000, people with chronic HCV infection had an uncertain fate with a reasonably high risk of dying early from cirrhosis and/or liver cancer. Today, an 8–12-week course of a daily oral antiviral drug results in a cure. In 2016, Dr. Sofia received the prestigious

Lasker-DeBakey Prize for his contributions to the discovery of the first HCV cure. There is currently no such cure for HBV.

The mission of the Hepatitis B Foundation is to find a cure and improve the lives of those affected by hepatitis B worldwide through research, outreach, and patient advocacy. If that wasn't ambitious enough, our tactical plan was to establish a research center, bring scientists and resources together under one roof, or at least in one place, and focus on discovering a cure. The model for me was the Manhattan Project, in which scientists were brought together to urgently work on developing the atomic bomb in the 1940s. That analogy got me into constant trouble, for obvious reasons. So, I modified the branding of the model to be the less draconian "NASA Moon Shot" in which scientists were concentrated on getting a man on the moon.

The principle, common to both the Moon Shot and the Manhattan Project, was to bring experts together and focus on a single goal. Of course, the resources, finances, and other components needed for us to achieve our goal were enormous. In the tens of millions of dollars, maybe more, depending on the definition of "cure" discovery. I knew that. However, it was important to have a mission that would excite and mobilize people.

Not everyone agreed with this approach. One strategy could have been the one used by the American Cancer Society (ACS) in which funds are raised from donors and then distributed to research project applicants. The applicants are typically academic scientists submitting research proposals to the ACS. In this model, the funding agency does not carry out or direct research on its own. Our feeling was that although this might have the virtue of supporting diverse research, it was fragmented, and we wouldn't be able to make enough impact with the dollars we raised. Our funding might be diluted by the other interests of a funded lab. Although this model has been extremely effective, our goal was precise—a drug—and we wanted to be more direct.

Along those lines, our champion Dr. Blumberg challenged the idea of aiming directly at our cure target. He thought it might be a mistake to focus so narrowly. Dr. Blumberg was very much a Renaissance scientist. He was a physician, biologist, anthropologist, naturalist, and philosopher. One Saturday, he showed up at our home and asked if my son and I had ever been to the New Jersey Pine Barrens, which was 30 miles away. We had not. He insisted we take a drive there, and then we spent the day with him as he identified birds, wild rodents, and God knows what else, and related it all to the area's natural history. He was a big thinker who saw and understood almost everything. To find a cure for HBV, he believed we should look at everything and would probably need to take an indirect path. Study everything. His analogy was that when shooting an arrow, you don't aim directly at the target. I answered that we wanted to fire a bullet. The HBF needed a cannon. I know, the paramilitary metaphors again.

Creating a nonprofit organization with Paul and Jan Witte wasn't too difficult—legally that is. Incorporation as a nonprofit 501c3, with bylaws and a small bank account, involves some intimidating paperwork, but nothing we could not handle. The designation 501c3 refers to the section in the U.S. tax codes that describes a type of nonprofit that can receive donations that are tax-free. Anyone can form a 501c3. And we did. However, funding research in a meaningful way is another story.

In 1991, we had nothing near the amount of money needed to support any programmatic activity, let alone an entire research institute. The biotech center was not even contemplated at that time. We opened a small storefront for the HBF office, and through my faculty position and lab at TJU, I was able to receive several research grants. Creating a research facility and starting research for the HBF was another matter. I needed to leverage what little money I could control (i.e., research funding for my lab at TJU) and align the interests of several institutions that, well, on the face of it, did not seem aligned. The idea was for the HBF to have a research lab at, and in partnership with, TJU, but located off the main campus. I wanted a "Leafy Liberal Eden" for research, outside the hustle and bustle of Philadelphia, where people would focus on hepatitis B. TJU needed to be convinced that this was a good idea. But that wasn't enough. TJU also needed to be convinced that the off-site location was a good idea. That still wasn't all. The institution outside the city *was* a good idea, and it should be done as an HBF-TJU partnership. The location outside the city was now going to need a third partner.

As explained a bit later, this third institution, the Leafy Liberal Eden, which we identified as the host for our partnership, turned out to be Delaware Valley College (Del Val). It was a small agricultural college in Doylestown, PA, approximately 30 miles north of TJU's campus. I had to convince leadership and staff at all three organizations of the value of a partnership. In 1997, it all came together, and Del Val agreed to build a new building that would house the new Jefferson Center for Translational Research and the HBF in rented space. Remember, the Pennsylvania Biotechnology Center, the hero in this story, did not yet exist. The partnership was ultimately the result of a forced marriage, you could say, between these multiple institutions. Three parents. Modern times.

1.2 Entrepreneurial Philanthropy

Raising money for hepatitis B research and the HBF was not easy. The traditional approaches were to (a) raise money to fund research at the nonprofit institution; examples include research universities such as Yale U. in New Haven, CT, or research hospitals such as St. Jude's in Memphis, TN, or (b) raise money to support research at selected institutions through existing nonprofit organizations (i.e., American Cancer Society, American Liver Foundation). We were thinking the HBF would be more like the (b) model.

However, in the early 2000s (a decade after we started the HBF), the Cystic Fibrosis Foundation took what I believe to be a revolutionary approach for a nonprofit pursuing a new therapy. Cystic fibrosis (CF) is a severe, ultimately lethal, genetic disease caused by a mutation in the cystic fibrosis transmembrane conductance receptor (CFTR) protein. People born with defective CFTR genes in both alleles usually do not live past their twenties without intervention. It is an uncommon disease, with 30,000 cases in the U.S. and perhaps 70,000 worldwide. It is, thus, considered an orphan disease. The U.S. FDA designates "rare diseases" that occur in fewer than 200,000 people. Medicines developed for rare diseases are called "orphan drugs" and may receive regulatory concessions. Despite regulatory concessions, big pharma usually has little interest. This, in a sense, justifies the orphan drug program. Although a fascinating scientific problem, the CF Foundation did not want to wait for someone or some place to become fascinated with the scientific puzzle to discover a cure.

In 2000, the CF Foundation invested $40 million in a small startup company (Aurora, which became Vertex) to fund the discovery and development of new drugs for cystic fibrosis. The CF Foundation retained a stake in the company. Vertex has successfully brought several new CF drugs to market, such as the breakthrough pharmacophore Kalydeco®. Today, these new drugs are essential for treating CF and extending life. People with CF can now expect to live in their 50s and hopefully longer. A major advance and a set of drugs that act in a new way by somehow restoring function to defective CFTRs have been achieved. Although by no means has a cure been found, the CF Foundation has done an admirable job serving its constituency while continuing its work with greatly enhanced resources. This model also allowed for breakthrough benefits for the foundation itself, and in 2014, it sold its rights to Royalty Pharma for $3.3 billion. The CF Foundation continues to be a major source of investment philanthropy for new cure research.

I liked this approach. However, we did not have funds for this model. We needed to be something in between a straight brick-and-mortar research institute and an investor. Experiences from our beginnings and then at the University of Oxford would ultimately suggest the path that led to the Pennsylvania Biotechnology Center.

Initially, the HBF focused on taking telephone calls from people with or concerned about HBV and traveling to our state capitals and Washington, DC, to advocate for more HBV research funding and use of the vaccine. We learned that there was considerably more interest and need than we had anticipated. We were initially fielding up to 30 calls a week, and the patient demands mushroomed after we went online, with the help of my computer and my Internet-savvy brother-in-law, Tony Oppenheim. Our website, hepb.org, went "live" in 1996, so we became pioneers on the Internet. The first server was the squeaky, dial-up AOL system that sucked down all of our bandwidth. By 2022, the HBF website has received more than three million unique visits per year, with more than 30,000 email and social media questions answered annually. Today, HBF has indeed become the world's leading, trusted portal of entry for information about hepatitis B and its related diseases, hepatitis D coinfections, and liver cancer.

In the beginning, during the 1990s, the HBF was truly a family affair. In addition to the help from my brother-in-law Tony, my sister Jamie helped Jan Witte with our newsletter, Joan answered phone calls and emails from anxious patients, my father-in-law served on the board, and my brothers, Tommy and Teddy, provided hospitality at our fundraising events (at the time they owned a chocolate and ice cream shop).

I didn't, however, just want to tell people about HBV. I wanted to *do* something about it. By that, I mean I wanted to become an HBV scientist myself and create the research facility that would work on the cure. Although I was a virologist when we started the HBF, I was not working on hepatitis B. To learn more about HBV and begin to work on a cure myself, I took a sabbatical in 1992 at the University of Oxford to work with Dr. Blumberg in Prof. Raymond Dwek's Glycobiology Institute. During that time, I learned both science and entrepreneurship. The research was fun and the caliber of scholarship was inspiring. That experience had an enormous influence on me, setting me on a course of research that endured throughout the rest of my career.

I studied the "glycobiology" of hepatitis B, a word coined by Prof. Dwek that refers to the study of sugars, also called glycans, which, when attached to proteins (i.e., glycoproteins), can alter protein structure, half-lives, and function. Glycan can attach to proteins in an "N" (to asparagine) or "O" (to serine or threonine). At Oxford, we determined that subsets of HBV glycoproteins (sugar-containing proteins) had strict requirements for N-glycan processing for their folding and function. This was a nice, solid, scientific finding, but unlikely to result in the kind of transformational result that would be a cure. I was also looking at Prof. Dwek's entrepreneurship. He had created the Oxford Glycobiology Institute, a largely translational research division in the Biochemistry department of the University of Oxford, as well as the first Oxford University spinout company, Oxford GlycoSystems. Translational research refers to research that is translated from discovery into practical commercial use. I took careful notice of both Prof. Dwek's new division and the spinout company.

Prof. Dwek had managed to thread the 1,000-year-old narrow needle of Oxford's traditions. He was carefully pulling drug discovery right through the small loop on that sharp needle of very conservative rules and attitudes. His institute was studying basic and applied biochemistry but also identified new drugs for viral infections and genetic diseases. Importantly, their work took one of their discoveries all the way through human use. The story of Zavesca®, which is still the only oral treatment for Gaucher disease, is an example of both the ability and willingness of a nonprofit institute to recognize a discovery of medical and commercial value, as well as an example of tremendous insightfulness and deduction. This drug is the result of Prof. Dwek and his colleagues Fran Platt, Ph.D., and Terry Butters, Ph.D., realizing the enormous potential of an unexpected result.

Gaucher disease is a recessive hereditary, lysosomal storage disorder, analogous to Tay-Sachs. It affects only a small number of children. It is the result of inheriting mutant GBA genes that fail to produce or specify defective glucosylceramidase enzymes, resulting in toxic accumulation of a set of sphingolipids called glucocerebrosides. In Prof. Dwek's lab, we were studying a sugar called N-butyl-deoxynojirimycin (NB-DNJ). My interest was that it could inhibit the maturation of HBV glycoproteins and might have value in treating HBV infection. Prof. Dwek, Dr. Platt (now Prof. Platt), and Dr. Butters observed that NB-DNJ caused a reduction in the accumulation of toxic glucocerebrosides in lysosomes. They then realized that NB-DNJ might be a good way of treating Gaucher disease since it would prevent the buildup of the toxic product causing the problems. They called this approach substrate deprivation. Their idea has worked, and their drug is a form of

NB-DNJ, which is now a life-extending medicine for children with Gaucher disease, and perhaps other lysosomal storage disorders. This is a remarkable story of an alert eye in the laboratory, knowledge of the biochemistry of a disease, and the mechanisms in place to take the discovery into human trials.

Like cystic fibrosis, Gaucher is an example of the U.S. FDA-designated "orphan disease," which, as we have seen, is largely ignored by large pharmaceuticals. Although HBV affects a much larger number of people than did Gaucher, the indifference of large pharma was strikingly similar. By the way, this is not to be critical or bitter about a market-driven industry such as pharma. Market-driven product development has delivered innovations that help define civilization. It works. That said, it does highlight the role and need of nonprofits, government, and universities in providing the research safety net for diseases that are not immediately attractive to large pharma but still of great unmet need.

The University of Oxford Glycobiology Institute had a favorable predisposition to drug commercialization rather than typical academic suspicion of entrepreneurship. Prof. Dwek embraced applied biology, which brought a drug from discovery all the way to human use. This caught my attention. The "academic to entrepreneur" approach was just emerging on university campuses in the U.S. It's somewhat ironic that it took a sabbatical in England to get me to see such an example of academic entrepreneurship. Even today, the UK is considered much less entrepreneurial than the U.S., ranking several countries behind the United States in the most recent ratings reports (GEDI, CEOWORLD 2021). However, Prof. Dwek had created a hotbed of innovation and translation. In 1993, I went home to America, excited.

I returned to my lab at TJU, re-energized and refocused on the study of HBV, as well as thinking more and more about creating a dedicated translational research institute for HBV, akin to Prof. Dwek's Glycobiology Institute. At that time, my lab was a modest operation. It had also suffered from neglect over the year of my absence. I had excellent young trainees, but they were a bit resentful of my absence and even more disturbed about the change in research direction. We had previously been in a lab studying only herpesviruses. To be fair, they had been left alone for an entire year. For me to return and start by announcing a change in our research direction was more upsetting than I had thought it would be. In the early 1990s, there was no Zoom or other online communications platform. Telephone calls were the only way for me to connect with the TJU lab when I was in England, and that was expensive. During my absence, the lab had to be fairly independent. Fortunately, one of the young people fresh out of undergraduate school, Anand Mehta, embraced my new energy and vision. He gladly took on the HBV work.

Even after my sabbatical year, I made several trips back to Oxford to continue work with Profs. Dwek and Blumberg, and Anand Mehta would often accompany me. This is an excellent example of the benefits of nurturing a young person from your group and putting their interests (to some extent) ahead of your own immediate self-interest. Yes, since Anand was among the sharpest people with whom I have ever worked, it would have been great, if not self-serving, for me to pressure him to stay with me at TJU to get his advanced degree. It was something of a sacrifice that I encouraged him to seek a doctorate with Prof. Dwek at the University of Oxford.

I am not saying it was a sacrifice of saints, and it is certainly what a professor must do for his students. Nevertheless, I would have loved for him to stay with me.

It worked out. After getting his doctorate at Oxford with Prof. Dwek, Anand returned to work with me in the U.S. for more than 15 years. He is now a distinguished (endowed) professor of considerable accomplishment at the Medical University of South Carolina. Along the way, he played a major role in our translational work in identifying cancer biomarkers, the development of our biotech center in Doylestown, and a good part in my professional development. I like this part of my story because it shows how nurturing relationships and encouraging trainee growth can result in lifelong relationships that are beneficial to the individuals involved and to their institutions.

Although Anand would eventually go to Oxford, I was able to bring a scientist in the other direction, from Europe, to our lab. In 1993, I brought a mid-career scientist I had met during a month's stay in Prof. Wolfram Gerlich's lab in Giessen, Germany. Prof. Gerlich was a leading HBV scientist, having done much of the early definitive work on viral proteins. Xuanyong Lu, Ph.D., was his research associate, with a temporary scientist status from China. He had apparently developed a way to infect tissue culture cells with HBV, a technique that had eluded scientists since HBV discovery. I know Dr. Blumberg was very proud that they were able to identify HBV without ever having to grow the virus in culture. That inability, though, frustrated continuing basic research on the virus. Dr. Lu appeared to have a method. Although that method never truly panned out, the results were real. Ultimately, the method proved to be too complicated, and better methods to infect tissue culture with HBV emerged.

Since Dr. Lu needed to leave Germany because of his immigration status, we were able to offer him a position in my lab at TJU. Funding came, in part, with a grant from The Blanche and Irving Laurie Foundation, championed by Harvey Rich, one of their Trustees. Xuanyong and his wife and two young boys came to the U.S., living with my family for about 8 weeks while we helped them find a home and adjust. They all spoke fluent German, some English, and, of course, Mandarin Chinese, so they seemed like geniuses to me. Xuanyong was a creative scientist who stayed with me for decades. Eventually, in 2015, he spun out his own company (Imcare) based on discoveries of biomarkers of liver disease made with me. Xuanyong was about as basic a scientist as I ever met, but he saw the potential in a finding we made in 2010 and took advantage of the new entrepreneurial environment in which he found himself.

My only regret is that our nonprofit organizations (HBF and the Blumberg Institute) have no financial stake in our spinouts. In addition, neither do I. There is no bad feeling between the nonprofits and the entrepreneurs, but I think this situation is unfortunate. This omission, to my mind, has to do with a combination of unclear policies and differing opinions of value. Needless to say, tightening policies that reduce misunderstandings and ambiguities is certainly an important part of building a sustainable academic business incubator.

1.3 A Leafy Liberal Eden

I returned from England with not only new vigor but also renewed vows to build a research center for HBV. Seeing the examples of translational research so successfully developed at the University of Oxford by Prof. Dwek also made an impression. By the mid-1990s, there were still no real HBF research labs anywhere. I was still on the faculty at TJU but was getting restless. However, I was in good shape research funding wise, with three federal grants (two from the NIH and one from the U.S. Dept. of Agriculture [USDA]). I had some scientific momentum, the prestige of research experience at Oxford, and the partnerships of renowned scientists such as Prof. Dwek and Dr. Blumberg that provided a bit of star power and enhanced credibility. Feeling empowered, I thought it was the right time to strike out on my own, leveraging that research and support to start a new institution.

The idea for the HBF to build a translational research institute focused on HBV on a college campus was now more pressing for me. A specific plan was taking shape in my mind. We would leverage my research funded projects (NIH and USDA grants) and recruit a number of world-class scientists who were also good teachers. Parenthetically, Dr. Blumberg had returned permanently to the U.S. He was willing to join us, although this would be on a part-time basis since he had other commitments at NASA and Fox Chase Cancer Center in Philadelphia. His involvement would allow us to brag about having a Nobel Laureate on the campus to which we chose to relocate. The college where we would locate and rent space would be chosen for its overall compatibility. We would benefit from the intellectual environment of a college campus, and the college would benefit from having great scientific research conducted on their campus. The collaborations and learning experiences that would also become available to their students and faculty would be at no cost to them. That was the value proposition to whichever host college would take us.

The hunt for a leafy liberal Eden was on. We didn't want to be in an urban area. For a variety of reasons, I wanted to stay within greater Philadelphia, so that didn't leave many options. The University of Delaware (U. Del) in Newark, DE, was high on my list of favorites. It had an excellent undergraduate science program and faculty, a handsome campus not far from Philadelphia, and realistic ambitions for strong graduate life science research. It even had a growing appetite for translational research. U. Del had offered me a position at the time I was looking for my first faculty appointment, so I had a soft spot for them. Although tempted, I favored an offer from TJU because of their virology program and proximity to U. Penn/Wistar, where I took an adjunct appointment and worked closely with a senior scientist, Nigel Fraser, Ph.D. Fortunately, there didn't seem to be any hard feelings.

I still had great affection for U. Del and knew several of the faculty leaders. At the time, they were considering big investments in biotech, even looking into developing a site across the street for an automobile manufacturer teetering on closure. Ultimately, they built a major facility on that site. In the early 1990s, I visited their campus with the idea of an HBF-led research center on their campus and met with the Dean and the Provost. It looked as if there was quite a bit of interest.

1.3 A Leafy Liberal Eden

While U. Del was pondering our pitch, my mother, Joyce Block, suggested I contact Joshua Feldstein, Ph.D., President of Delaware Valley College (Del Val). It is ironic and confusing that U. Del and Del Val have similar names, but they are completely different schools. Del Val is a small agricultural school in Doylestown. It has since gained university status but is still fairly small, and back then, it was even smaller. My mother, always trying to keep her six kids and their families close together, saw an opportunity to bring us back to Bucks County, PA. That's where she and my father and the other five kids and their families lived, having moved from Buffalo, NY, in the 1970s. Bucks County wasn't an entirely crazy idea. It was (and is) the "bedroom" for many biotech and pharma scientists who worked in nearby counties. I had already been looking up that way. I hadn't seriously considered Del Val, but we had visited a couple of factories that were closing. We even got a tour of a Navy Air Base in Warminster, Bucks County, which was scheduled for mothballing. Dr. Blumberg, however, nixed the idea of not being on a college campus, and I agreed.

Unlike U. Del., Del Val was tiny and had little tradition of professional scientific research. Nevertheless, it had a pleasant, if modest, campus right on the outskirts of Doylestown, PA, which is one of the most charming small towns in the U.S. In addition, of course, it was near my family. My mother gave me Dr. Feldstein's office telephone number and threatened to make the call if I didn't. Having a bit of pride, I made the call.

Dr. Feldstein saw me almost immediately. He then joined me in imagining what could be done and all of the possibilities of having a medical school satellite operation and a nonprofit research foundation on his campus. For me as a young faculty member, this was a heady experience. The president of this college was encouraging me, even urging me, to bring my research program and idea for a translational institute with our very small foundation to his campus. The deal was that Del Val would build us a building that we would rent and in which we would locate our foundation and research team. He argued that this should be done in partnership with my employer, TJU. He reasoned I would not, should not, give up a tenured faculty position (which I was willing to do) for this speculative effort. Moreover, having TJU involved would add credibility and, frankly, security to the project. I agreed, when in fact I had already been pitching the idea to the TJU Dean, Joseph Gonnella, M.D.

The proposal to Dr. Gonnella was that TJU would sign a master lease to construct a new lab and instructional building to be called "The Jefferson Center" on the Del Val campus. HBF would have a sublease and operate in partnership with the TJU scientists, which included me and my relatively small but well-funded group, as well as labs from Hilary Koprowski, M.D., a world famous, albeit controversial, scientist. Before coming to TJU, Dr. Koprowski was director of the Wistar Institute, a prestigious research nonprofit on the campus of the University of Pennsylvania. Dr. Koprowski was considered the "father" of the rabies and first polio vaccines, and his lab at TJU was producing a recombinant HBV vaccine in plants. The idea, which was extremely novel, was that a plant-produced vaccine could be grown in countries that did not have access to industrial-scale facilities to produce the standard vaccine. The idea was clever, although there were a few other labs trying similar approaches.

The work was a good fit for us, and since his researchers needed agricultural facilities, notably biohazard-restricted greenhouses, these could readily be built at Del Val. It helped in my justification of the Del Val project to TJU's leadership.

Dr. Gonnella became our champion on the TJU board, and Dr. Feldstein became our champion with the Del Val leadership. TJU was supportive of the HBF, but that wouldn't be enough for them to foot the HBF research bill. The funding from the research grants I received as a professor at TJU played a critical role in making the deal between Del Val and TJU possible. And in enabling our operation once we were at Del Val. Those grants came with considerable "indirect cost recovery." Indirect costs are paid to the host institution (my employer TJU at the time) by the sponsor (the NIH, my primary source of research grants). Indirect costs pay for the research overhead expenses of lab rents, utilities, and non-scientific administration. I made the case to TJU that those grants would provide all the funds needed for the basic research as well as overhead of lab rent for the new building on the campus of Del Val.

Since my work was, at that time, largely translational antiviral and HBV research, this was a good fit for the HBF. It was easy to justify locating the work off the TJU campus by making an argument about keeping their academic research uncontaminated by applied research and that translational work would best be done at an off-campus site. The TJU Dean, Dr. Gonnella, was supportive of the HBF and of what I was trying to do. He got behind the project and was our advocate to the TJU leadership in creating a satellite operation named the "Jefferson Center for Translational Research," in which my labs would conduct translational research on HBV. It would be a TJU division located in an off-site building being constructed for us on the campus of Del Val and in partnership with the HBF.

The agreement was signed in 1997 by the presidents of both schools (Paul Brucker, M.D., for TJU, and Joshua Feldstein, Ph.D., for Del Val). Dr. Koprowski's plant vaccine group and my lab, along with a newly declared HBF office and lab, moved from Philadelphia to Doylestown into our new building on the Del Val campus in 1998. A few students and one other professor from TJU with translational research projects joined us within the year as well, bringing our head count close to 15 people. Our small pioneering group occupied ~14,000 ft^2 of a smart, new 28,000 ft^2 building. This space sounds a bit more grand than it was. We had a net of perhaps 4,000 ft^2 of labs, with the rest of the space being hallways, offices, and classrooms. There was also 2,000 ft^2 of space for animals, which Del Val operated, and it was not for our use. That said, the creation of the Jefferson Center on Del Val's campus was a major inflection point for the HBF, which could now brag that it actually had a research center. We would operate as a satellite of TJU, focused on HBV.

Since the construction of the new building at Del Val and our work would be under the banner of a partnership with the HBF, our goal of creating a bricks-and-mortar place became a physical reality. The foundation was previously mostly virtual, initially operating from our homes and then a storefront office. The project was to be financially justified by my research grants and the indirect costs they paid. The "Master Lease" called for TJU to pay rent to Del Val for the use of 14,000 ft^2 of a

28,000 ft^2 new building. The HBF would be a subtenant. The money to do this at TJU would come from research grants for my work, as well as whatever funds the HBF could raise.

This wasn't as risky as it sounds since I already had several research grants that came with indirect costs to pay for administration and lab facilities. Del Val would charge TJU a fixed fee for building use, so TJU knew in advance what their expenses would be. It was fairly straight forward to calculate the total dollar amount of the indirect costs. TJU signed a 10-year lease with Del Val, to coincide with Del Val's borrowing. Of course, the research funding being brought in was only assured for a few years of a 10-year lease. TJU certainly had some risks. Not to mention the reputational risks of a satellite campus working on HBV translational research on the campus of a small agricultural college, 35 miles away from the main TJU campus.

Del Val president Dr. Feldstein, wisely, in my opinion, embraced the project and was enthusiastic. And, why not? At no cost to his small college a major U.S. medical school was offering to partner with a nonprofit foundation and locate a group of highly accomplished scientists and scholars with international reputations on his campus. Several of the scientists who would locate on the campus from TJU were funded by the NIH and USDA. In addition, TJU's rent would pay for the new building. Instantly, the Del Val faculty and students would have a world-class research operation and access to a major medical school on their campus. This was an unprecedented opportunity for a small college. I cannot think of any other small undergraduate college in the U.S., including the most selective and prestigious, that had such an arrangement.

TJU and the HBF were bringing to the Del Val campus a research portfolio, a roster of scholarly seminars with visits from thought-leading researchers from all over the world, and full-time scientific colleagues onsite. Some of these scientists included members of the U.S. National Academies, the Royal Academy of Medicine, Lasker Prize winners, and even a Nobel Prize winner, Dr. Baruch S. Blumberg, who was scheduled to visit regularly as part of the HBF. We had many VIP luminaries in science and medicine working with us and visiting. What an asset for a small college, opportunity, and, frankly, competitive advantage.

This partnership should have been regarded as a great deal for Del Val to be enthusiastic about. And, to Dr. Feldstein's credit and vision, he was. However, he retired, as he had planned before we had even been there a year. His successor, Thomas Leamer, Ph.D., was also supportive, but we faced an unexpected surprise— a strangely hostile Del Val faculty.

There was apparently resentment among Del Val faculty toward those of us from TJU. First, the building costs exceeded the estimates, but the TJU payments to Del Val were fixed. I still thought Del Val was getting a good deal, but the math had changed, and thus, presumably, so did the net financial benefits to Del Val. However, we were all still coming: the scientists, their research, just as famous and accomplished as I promised. How can you put a price on that? Well, I guess they did. I came to realize (or was told) that the TJU payments could no longer cover the expenses of the entire new building, but rather only our 14,000 ft^2. That didn't go over well with our hosts.

There was friction with the Del Val faculty that went beyond the financials. Rather than being received as stimulating and worthy new colleagues, many Del Val faculty appeared to be threatened by us. Frankly, it was difficult to tactfully provide reassurance that we were not interested in taking their jobs without being insulting. Perhaps it was naive to have assumed that our arrival would be perceived by them as a great complement. We were primarily researchers (which Del Val didn't have much of), and they were primarily undergraduate educators (which we did little of). It was true that TJU and our collaborating scientists working on the Del Val campus were all paid on a medical school scale, and although I don't think we were arrogant or pompous, who knows?

Our cultures were very different, and I underestimated that impact. Del Val was and is still a small undergraduate teaching institution with approximately 100 faculty members and 1,700 undergraduates. It is science and career-oriented, and we hoped this would strengthen the relationship between them. Del Val had ambitions of becoming a university, which it was ultimately able to accomplish, and this should have increased our appeal to them. However, the differences between our cultures proved too great. We were trying to operate a research facility with full-time staff that made transformational contributions to medicine. Their faculty generally worked on a 9-month calendar and were dedicated to teaching undergraduates. Their interest in research was confined mostly to the extent that it could have a teaching benefit for their students. There's absolutely nothing wrong in that, but I thought that would be all the more reason for cooperation. However, many apparently thought we were there to take their jobs. Air was even let out of the tires of our cars multiple times. It was hard to believe.

Even with the occasional flat tire, we persevered and thrived. The research progress at the Jefferson Center would ultimately give us the credibility and means to leave Del Val and pursue our first experiences with startup company creation. We were able to create something of a world of our own in which applied biology and development was emphasized. We generated the technology that led to two faculty spinouts, Kimerigene and Synergy Pharmaceuticals.

We developed a durable, reproducible, scientific approach. To discover new drugs for HBV, we designed "screens" and obtained a "compound library." These libraries are collections of compounds (molecules) that have drug-like potential or can lead to drugs. Such collections are usually the crown jewels of a pharmaceutical firm. Screens, a method to detect a drug, are produced, and then, one by one, each compound in the library is tested for activity. Big pharma companies typically have more than one million compounds in their library. To my knowledge, there were no examples of nonprofits or even universities with many, if any, compound libraries at that time. We needed one.

The company ViroPharma, which was developing antivirals for the common cold, located in Malvern, PA, had some of the most talented scientists in the area. It was, however, on the verge of collapse and conducting a fire sale of its assets. In those days, early 2000s, compound libraries were not readily commercially available (or were difficult to acquire), and to my knowledge, there were few nonprofit organizations that had their own. ViroPharma had one of the best. Not large, but, given the brains at ViroPharma, I assumed a very well-curated library. I approached them about buying part of their library (~80,000 compounds), as well as some of their screening robots. In part with a donation from Catharine and Rob Williams and Catharine's family foundation, The Gunst Foundationir, we were able to buy a part of their library, liquid handling robots, and hire a couple of their scientists. With that, the HBF research labs instantly acquired a capability that few other nonprofit research organizations had. We set up the appropriate assays, had professional industry-experienced scientists, and began to conduct high-throughput screens to test for compounds active against HBV.

With the compound library, we screened and identified first-in-class small molecules that inhibited the HBV "s" antigen (the surface or viral coat protein) and HBV capsid formation in tissue culture. These ultimately led to the creation of the HBV startup Enantigen, later acquired by OnCore Biopharma. Tekmira then merged with OnCore to become the publicly traded company Arbutus Biopharma, which has clinical phase HBV compounds. We also spun out Synergy Pharmaceuticals, which became publicly traded. It developed and received FDA approval for the drug, Trulance®, indicated for digestive distress, and was invented by onsite scientist Kunwar Shailubhai, Ph.D., M.B.A., with CEO Gary Jacob, Ph.D.

While still on the Del Val campus, two scientists from Wyeth Lederle, Satish Chandran, Ph.D., and Cathy Pachuk, Ph.D., came to us and developed what turned out to be a revolutionary approach to treating HBV and other diseases through a program I called "Out and Up." Briefly, this is where we offered exits from big pharma companies for mid-career scientists who were frustrated with their situations. They had ideas their companies did not want to pursue but in which we had an interest. In those days, in the late 1990s, pharma did not readily downsize scientists. I know this sounds almost quaint by today's standards. However, this meant there might be scientists a company would value, but then stick them with golden handcuffs. Or the companies liked them enough to not fire them but instead move them to research the scientists might not want to do. These scientists, we reasoned, might be willing to exit from their big pharma jobs, and come to us, under the right circumstances. They were, of course, paid much more than we could pay. Therefore, I proposed our program in which they could leave their positions at big pharma, ideally with a "grant" from their company, to carry out the new ideas they had at our institute. The big pharma could retain rights to the new work or not, depending on the arrangement. Remarkably, sometimes these big companies did not wish to retain rights and were just happy to facilitate amicable exits.

I had met Satish and Cathy and learned of what they were working on, and it sounded as if it could have application to HBV. They were among the first to see that silencing RNA (siRNA) could be used as a therapeutic. These siRNAs, as they are now called, have become one of the most promising new therapeutic approaches for HBV and other disorders. The discovery of siRNA, in 1997 by Mellor and Fire, resulted in a Nobel Prize. Satish and Cathy were very early into this, and their 1999 patents, which they used to create the company Nucleonics, may be the first human-use patents for this technology. Nucleonics was acquired by Alnylam, which is currently a leading developer of siRNA therapy and, as of 2022, has at least one

FDA-approved drug. The value of siRNA is now widely recognized. The companies Arbutus, Roche, Johnson & Johnson, and others have or have had human clinical phase siRNA programs for HBV.

With these accomplishments, we had bragging rights. Things were cooking at the Jefferson Center on the campus of Del Val, and our Out and Up program was beginning to be recognized. We tried to position ourselves as a little but mighty academic research powerhouse that could spin out new technologies. The only one in our area. In addition, we made sure we got noticed. We made sure our Congressman, Jim Greenwood, who was a fan and early supporter, knew about us. Our PA state senator, Joe Conti, adopted us as well. He saw what we could do and introduced legislation that provided a path for us to receive support for a biotechnology center.

1.4 9/11 and Creation of the Pennsylvania Biotechnology Center

Even with our successes at the Jefferson Center and HBF, it was still very unlikely that we (meaning the HBF) could afford to hire the talent needed to make much of a dent in the problem of hepatitis B. We would need scientists and other professionals representing many different types of expertise: chemists, biologists, virologists, computational experts, drug discovery and drug development professionals, businesspeople, and so on.

We were still in rented space on the campus of Del Val, but the idea of a biotech business incubator was becoming more of a necessity and more irresistible. Based on examples of pharma and research institutes I was modeling, we would need to become a mini-biotech campus of at least 200–300 people in a facility more than 100,000 ft^2.

The funds needed to create this research utopia were nowhere in sight for us. The seemingly instant success in raising money and hiring great talent with the startups we had created had an impact on me. It gave me an idea. These new companies (Newcos) would, at launch, typically raise several million dollars, which was enough to hire 5–10 excellent scientists, rent decent lab space, and get work done without a great deal of fuss. They would necessarily have to locate themselves somewhere else. There was no room for them to be at our facility and no appropriate mechanism to locate down the hall from us, even if they wanted to do so. They were for-profit, and we were nonprofit. Therefore, rather than being down the hall or even on the same campus where we and they could benefit from each other, they were in another town or even in another state, despite having a shared mission and complementary skills. It seemed to me that by keeping us all together, we would all benefit, and this would provide a strategy whereby we could achieve the human resource and physical growth we needed.

I realized that we needed to create a biotechnology life sciences incubator in which our faculty-created spinout Newcos could locate. In addition, we would attract like-minded academics from other institutions to spin out their companies. We would also make a pitch to those in pharma with entrepreneurial ideas, such as Satish and Cathy, to either partner with their pharma employers in spinning out a Newco or take the leap on their own (with our help). One more thing. The business incubator would be owned and managed by the nonprofit, the HBF. At this time, our research organization, Baruch S. Blumberg (the Blumberg), did not yet exist.

Lest I forget to mention, we would need to be sensitive to our hosts, Del Val, who had agreed to our relatively small operation on their campus and not necessarily to having a biotech Disneyland carved out of their farmland. I proposed a deal.

We (meaning me representing the HBF) would write a grant to the state of Pennsylvania to build a biotech incubator that the HBF and Del Val would own, 50-50, but the HBF would manage. In addition, the HBF would be a senior partner, appointing the incubator's president. I had also offered to include TJU since I was still a professor there as a partner. They declined. TJU was willing to be a tenant but not an owner since they were probably concerned about liabilities. By this time, around the year 2000, Dr. Gonnella was stepping down as Dean of TJU, and the new administration was less enthusiastic about our operation so far from the main campus. TJU ultimately opted out. Dr. Feldstein (now retired but still an influential member of the Del Val board) and several other board members, along with their new president, were very supportive. Why shouldn't they be? I was proposing that 10 acres of their more than 600-acre campus be their contribution to the partnership in the creation of what I hoped would be a major biotech complex. The site chosen was at a remote part of their campus with access that was separate from their main entrance. This was so it wouldn't disturb the nature of their college campus and to give us more identity distinction. In exchange, Del Val would become home to world-class research, entrepreneurship, and opportunities of historic proportions. For HBF, this would represent exponential growth.

However, we needed money for this. The funding could come from private sources, and the biotech center would be a for-profit entity or from public sources and it would be a nonprofit entity. I was open to both possibilities. To hedge my bets, I wrote a grant to the Commonwealth of Pennsylvania. In my proposal, space was specifically carved out for the HBF and the Blumberg Institute.

I wrote the grant, sharing it with the new Del Val president, Dr. Leamer, for his approval and ideas. He had no significant suggestions but at least was supportive and encouraging. I asked for help from a local entrepreneur and land developer, Rick Lyons, who had developed an attractive business campus on the edge of town. I liked the way his business park looked: farmhouse-styled buildings that blended in with the rest of the town's architecture. I approached him cold with my idea, and he provided concept drawings by his architects. One possibility was that if we did not obtain grant funding, he would build the complex as an investment. There was no actual agreement, but he generously gave me the materials to be used in my proposal at no charge.

Sometime in 2000, I submitted the proposal to the state's redevelopment authority program, which required that the recipients match whatever money was granted, dollar for dollar. That caused me to greatly reduce the scope of the proposal from a grand $20 million plus complex to a more manageable $15.8 million plan. Of this amount, $7.9 million would be from the state, and $7.9 million would need to be the match from us. This was to be comprised of (a) $1.4 million appraised value of the 10 acres contributed by Del Val and (b) ~$6 million value of the new science building we were occupying on their campus. The balance (~$500,000) would be borrowed or contributed by the HBF and Del Val.

The proposal was sent off to Harrisburg, the state Capitol, and, being a bit naive, without informing any of our elected representatives. Our state senator, Joe Conti, apparently learned of the proposal and became a passionate, effective advocate, I was later told. The timing of my proposal turned out to be fortuitous, but on the heels of our national tragedy. The world changed.

On September 11, 2001, terrorists hijacked planes and flew them into the World Trade Center in New York City and the Pentagon near Washington, DC, as well as into the ground in central Pennsylvania. The nation mobilized, and a new cabinet-level department was created to address these threats to the country. PA Governor Tom Ridge was named by President George W. Bush to become the nation's first Secretary of Homeland Security. His departure meant that the new PA governor would be Mark Schweiker, who was the lieutenant governor.

Mark Schweiker had previously been Bucks County's commissioner and was well known to us and certainly to state Senator Conti. That's all I know, but presumably their friendship had a lot to do with what came next. To my surprise, in 2002, I received a call from Governor Schweiker's office and Senator Conti's offices telling me my proposal had been selected and would be funded in full. Although my grant wasn't in the correct format, I was told not to worry because successful applicants would receive assistance with this process. In January 2003, Gov. Schweiker came to the campus of Del Val and presented me, HBF president, and Dr. Leamer, Del Val president, a ceremonial check of $7.9 million written to the "Hepatitis B Foundation and Delaware Valley College."

Rob Loughery, then a young staffer from Congressman Jim Greenwood's office, helped me put together a proper application and then stayed with us as project manager for the center. It took 2 years from the time of the award notice to come to an agreement with Del Val about what should be built and start the project. Parenthetically, Rob would go on to become a Bucks County Commissioner several years later.

Del Val subjected me to their community process, in which I was to meet with various faculty members to hear their ideas about what should be built. It was exhausting. Their ideas included a museum, a library, animal care facilities, classrooms, and labs. There was some discretion in what could be done, but the deal was already sealed, and much of what the college faculty was suggesting would not have been allowed. The application had been submitted. The state funding was given to build a modest-sized (28,000 ft^2) biotechnology incubator with space for the HBF and our research organization to complement the existing 28,000 ft^2 building in which our 14,000 ft² footprint is currently located. I had set aside about $1 million of the $7.9 million for Del Val to repatriate the 14,000 ft² we would abandon, but beyond that, the deal was done. Needless to say, this community process did not improve relations with their faculty.

Grow or Go

Although the proposal to the state called for a new, 28,000 ft² lab and instructional building to be constructed in a 10-acre carve out of the college's land, we intended for this to be just the beginning of the ~120,000 ft² multi-building complex anticipated in the drawing from Rick Lyons. The word of this created new tension on campus.

To be fair, the scale of my ambitions for our research center could have understandably upset a small college if it did not feel aligned. The mission of the HBF was and still is to help find a cure for those living with chronic hepatitis B. I had been observing companies being spun out of our technologies and from other academics and then locating themselves distant from us and our work. Our research had already generated the technology that led to a couple of startup companies. It seemed to me that if these companies would locate near our labs, ideally on the same campus, if not in the same building, there would be synergy or at least substantial mutual benefit. I also appreciated that a university campus might be an inappropriate place from which commercial, for-profit entities should operate, given their mission of education. However, the HBF mission was translational. Through our research labs, we wanted, and frankly existed, to translate our discoveries into new drugs. Being co-localized with like-minded scientists and entrepreneurs, I reasoned, would complement and accelerate our mission.

It was becoming clear that the cultural environment of Del Val was not right for us, and the physical space was not enough. I declared a Grow or Go strategy, but since I did not think we could be sustainable at the size of staff or facility we were on the Del Val campus, the Go option was looking more likely. I knew we would need hundreds of scientists working in space many times what we were in at Del Val. If our 14,000 ft² footprint made faculty at Del Val uncomfortable, new proposals for expansion that I now had in mind could lead to even more open hostility. I would be the first person to agree that the larger proposal I had in mind could have a tremendous influence on Del Val's campus aesthetic, personality, and operation. Of course, in my opinion, all of that would be to the benefit of their campus. However, this was not a common position of their faculty, as far as I could discern.

One more thing. It was not helpful for us to have identity confusion between our institutions. Del Val was a solid regional undergraduate college. The HBF and the Blumberg Institute were increasingly recognized as national research organization. We needed alternatives. We needed to move out and up. An academic industrial revolution came to the rescue, providing the momentum we needed. An alternative that could relieve tension presented itself.

Doylestown is, and was in 2003, an extremely attractive town—smart, well-maintained Victorian-era homes with nicely landscaped yards. In addition, there are plenty of other amenities, including a wide variety of high-quality restaurants, museums, movie theaters, and fashionable shopping. It was a thriving, prosperous little town, which, as the county seat, assured a steady flow of consumers for local businesses. It was also in transition. There were a number of light manufacturing companies in the surrounding area that were in distress. Several were closing or on the verge of failure. The low-wage jobs were incompatible with the new, more upscale bedroom community of the pharmaceutical and financial industries that Doylestown had become. It was difficult to find people willing to work in those factories. One by one, they closed and were becoming useless eyesores.

Congressman Greenwood and his assistants Rob Loughery and Pete Krause suggested that we use the state money to carry out an adaptive reuse for those distressed manufacturing sites. Transforming an old factory or warehouse into a business incubator, they reasoned (and I agreed) was a much more civic-minded way to spend the money than digging up farmland on Del Val's campus. Congressman Greenwood said to me, and I pretty much am quoting him here, "If you do this, they will build a statue for you in this town."

Senator Conti then drove me around town, and we took roadside looks at sites I frankly had never previously noticed or been aware of. I particularly liked an old building in the borough, a few blocks from the town center, that had been an old pie factory. It was being used by the PA Department of Transportation. This would prove to be unobtainable. Rob and Pete drove me to an old print distribution facility, about 1 mile from the town center. This was a bit of an inconvenient location, too far from the train and bus stations for most people to walk, and on an obscure road with dated manufacturing buildings that gave it a very unappealing, grim, industrial look. Not truly what you expect from a modern biotech center. The DA Lewis Print Company, a print distribution center where more than 140 people had worked, sat on 10 acres. With a 40,000 ft^2 warehouse connected to a 20,000 ft^2 office building, the complex was for sale. Only a few office workers remained. Everyone else associated with the operation had either lost their jobs or had relocated to their new site in the south. I recall the asking price being about $3 million.

To consider this opportunity, we had to first convince Del Val to agree to have the biotech center located off their campus and receive permission from the state of Pennsylvania to allow for the grant goals to be achieved at a different location and in a different way (adaptive reuse rather than new construction). The state authorities were delighted to see us use the funds to bring value to a blight site, and the support of our congressman, county commissioners and state senator made this process easy. We especially had the enthusiasm and support of our County Commissioner Chair, Mike Fitzpatrick (who went on to become our Congressman, succeeding Jim Greenwood). Ironically, Del Val, who had been ambivalent about having us on their campus and who I thought would be eager to show us the door, caused multiple delays. I am not sure why, but Dr. Feldstein was a powerful advocate for us, and eventually the college relented.

TJU's new administration saw this as a chance to exit. If we were going to move, they wondered why we couldn't do everything we were doing back on their Philadelphia campus. We were well-funded, and I was a tenured professor, so this was not about threatening to get rid of us. Rather, I took this as the behavior of a

1.4 9/11 and Creation of the Pennsylvania Biotechnology Center

more conservative, more cautious administration that wanted to make better use of their urban campus. They even asked me if I would be interested in serving as interim chair of the Biochemistry department since there had recently been a vacancy. I declined, making it clear that my future was with our plans in Doylestown.

That meant I would be separating from TJU after more than 20 years on the faculty. That left open a couple of possibilities. In one scenario, I would take this as an opportunity to leave TJU for the HBF and its recently created translational research organization, now known as The Baruch S. Blumberg Institute (also called: The Blumberg Institute or BSBI). I had been a volunteer at the HBF since its inception, with my salary and my research support from TJU. Why not leave? I had lots of grant support, and if that could be transferred, the HBF would instantly be an NIH-funded institution. That would be a bit risky for me, and giving up a tenured professorship would be quite a sacrifice. On the other hand, I could try making a deal with another university. In this case, I could make a new arrangement, one in which they fully embraced the biotech incubator, the HBF, and everything we were trying to accomplish in Doylestown.

I pursued the second option, and in no time, Drexel University's very entrepreneurial president, Constantine Papadakis, Ph.D., declared himself to be fully supportive of us. Drexel had recently acquired Philadelphia's Hahnemann Medical School and the Medical College of Pennsylvania and was eager to find innovative ways to build their program. Drexel, similar to TJU, was located in Philadelphia but geographically closer to the University of Pennsylvania. It was highly regarded as a school for engineering and other applied sciences. Its acquisition of two historical Pennsylvania medical schools was something of a shotgun marriage. These schools had fallen into financial distress, and the state of Pennsylvania had intervened to help find and fund new relationships that would keep them in operation. The Medical College of Pennsylvania was founded in 1850 as a college for women's medical education. Hahnemann, founded in 1885, also had proud and successful alumni, many of whom practiced in Pennsylvania. Therefore, there were important reasons and powerful local voices to keep them in business. A compelling argument could be made that bringing a medical and engineering/applied biology-oriented university together made a lot of sense. Drexel's acquisition of my TJU division, even in Doylestown, appeared to be a good fit, too, with our translational research mission.

We were still well-funded for a small group and completely self-supporting. The Drexel vice provost, Bill Stephenson, Ph.D., came to Doylestown with his colleagues and pored over our books. We spoke about my vision, and before too long, he offered me a "celebrity professorship" (their term) with permanent tenure of salary and title. In the arrangement, they also offered faculty positions to the senior research staff of our division. It was a good deal for Drexel, with little risk. Not only were we self-supporting, we would be turning over a considerable annual fund surplus to Drexel. This surplus would be millions of dollars over the next several years. Therefore, with no investment in our programs, they acquired an excellent research team that was becoming among the most significant in the field of viral hepatitis, providing new research bragging rights for them as well as an affiliation with a nonprofit patient advocacy organization. We were also teachers for their students. It was good for us, too. We had the credibility and resources of a major university, new faculty positions for bright young researchers (who did not have these titles at TJU), and I received a much-appreciated raise in my salary.

Out went the Jefferson Center sign. In came the sign for the Drexel Institute for Biotechnology and Virology Research. This time it was understood and written into my employee agreement that I was a Drexel professor and serving as the onsite volunteer president of the HBF and the Blumberg Institute. Drexel would now have a formal partnership with the HBF. Moreover, the deal with Drexel was completed with the understanding that our group would be relocating to the new business incubator.

In 2006, we were still formally the Drexel Institute for Biotechnology and Virology Research. So, Drexel moved with us when we moved off the college campus to the newly renovated warehouse on the other side of town, which we named the Pennsylvania Biotechnology Center or PABC. The Drexel Institute was still a part of Drexel University, but the PABC, the renovated warehouse building, was a 50-50 partnership between HBF and Del Val. Ironically, after treating us with suspicion and hostility, many at the college objected to moving the research organization and the soon-to-be-built biotechnology center off their campus.

Approximately 17 of us, which included research assistants, students, and faculty, moved our labs and offices into the newly renovated space. The experience was exciting but also intimidating. 17 people. This number seemed especially small given the enormous amount of space into which we moved. Nevertheless, it was our space. Our building. There was 62,500 ft^2 of enclosed space. We were only using a few of the labs and offices. There was as much a feeling of vastness and emptiness as an opportunity.

It was a terrific feeling to have new labs and instructional space that we owned and that were created for us. The physical space felt fresh and new and showed little evidence of its warehouse past, thanks to the imagination of our architect, Steven Cohen of Princeton. I especially liked that my colleagues from TJU, now Drexel faculty, had faculty ranks and labs of their own. Of course, the HBF and the Blumberg Institute also had their own offices and labs, which was truly a big deal for me and the foundation.

The group of pioneering scientists that joined me at the new site were, in my opinion, as good as any in the country. Although, formally, we were united by my funding and overall coordination, these were independent scientists. They were all committed to studying HBV and liver cancer, making us a formidable, focused group of interconnected research teams. They would provide the seeds of several new inventions for the early detection of liver disease and new therapeutics for chronic hepatitis B. Many of these Blumberg Institute-based technologies would lead to new companies located at the PABC.

As mentioned before, these companies included Enantigen, OnCore (now Arbutus), ImCare, Synergy, and JBS, to name a few. Nucleonics, the company created by Satish and Cathy, who joined us in 2000, did not locate their company at the new biotech center because they needed space before our renovation was complete. In fact, their immediate demand for space was part of my argument to the state that we were spinning out new companies that needed more space. Instead, they hired one of our faculty members, Patrick Romano, Ph.D., and moved into their own new labs in Horsham, PA, approximately 7 miles from us. Nucleonics, as mentioned, was developing siRNA therapeutics for chronic hepatitis B, very much in keeping with our purpose.

Our partner in the PABC at that time, not to forget, was still Del Val. By agreement, they also had space at the new biotech center. They used classrooms at our new site and were able to take over the entire renovated space we abandoned on their campus. I had hoped they would feel more responsibility for us and provide security, and perhaps some maintenance and lawn services. For a fee, of course. However, they provided neither. The 17 of us sat in the enormous facility, alone. I watched the parking lot and road nervously from my office window, a bit worried about stragglers coming in, either just curious or malicious. We quickly hired a handyman, who also served as security, being the strongest person in the building. In addition, technically, we had all of the services that Drexel University could provide. And, slowly, but steadily, we grew. In unusual ways.

We could not afford to purchase high-tech equipment such as mass spectrometers, which we needed for our proteomic and chemistry work. We sought instead to recruit and nurture startups that had the capacity and equipment. Ramilla Philip, Ph.D., and her husband, Mohan Philip, Ph.D., joined us with their two-person startup, Immunotope. Their company had a method to identify antigen epitopes naturally presented by human T-lymphocytes, and this could be important to developing therapeutic vaccines for chronic hepatitis B as well as many cancers. Briefly, they could immunoprecipitate MHC (HLA) molecules complexed with antigenic peptide epitopes isolated from within an infected or cancer cell. They brought with them expertise and equipment I thought would be useful to Drexel/Blumberg scientists and the Center as a whole. We hired Ramilla as a member of our faculty, and she began to work on HBV, and Mohan, as our Keystone Innovation Zone (KIZ) coordinator. This is an example of how we leveraged our small organization to attract talented people and resources that we could not afford on our own but became valuable resources for the institution. Immunotope was eventually acquired by Emergex, and we eventually developed our own in-house proteomics capability.

KIZs were state-designated sites that conferred specific tax advantages to companies as well as eligibility for specific state grants. We received our designation and hired our second KIZ coordinator, Kathy Czupich, MBA, who was also our CFO. Kathy later stepped away from those positions to establish her own startup at the Center, Artemis, which was created partly at my urging and with a grant from the Blumberg Institute to provide administrative services to the Newcos I anticipated would be created at the Center. Artemis focused on helping with SBIR grants and is now a leader in this area, and has provided a valuable service to our Newcos.

We soon realized medicinal chemistry was needed. Most of us, as Drexel and Blumberg Institute scientists, were trained as life scientists (biologists, physicians, and virologists). This would only be half (or less than half) of an industry-grade therapeutic drug development team. Therefore, we were lucky that Alan Reitz, Ph.D., left his position at Johnson & Johnson (J&J) to create a company focused on offering medical chemistry design and synthesis services. We also made a deliberate outreach to bring Michael Xu, Ph.D., and his startup company to the Center. Michael was a medicinal chemist, formerly at J&J, and Pharmabridge was his medicinal chemistry company. Most of its talent was located in China. We hired Michael to be a member of the Blumberg's faculty and to lead the medicinal chemistry programs for one of my federal grants. Having industry-standard medicinal chemistry professionals at the Center and on our grants working onsite with us gave us a real competitive advantage relative to other academics. Our submissions had sophisticated chemistry proposals, and frankly, our progress was excellent. Michael and members of his small team, notably Yanming Du, Ph.D., were able to produce hundreds of chemical modifications of "hits" (a compound from a molecular library that showed activity or positivity in an assay) that would come from screening our compound libraries. This allowed us to show funders and investors that we could carry out rational drug designs that approximated what big pharma was doing. Parenthetically, Michael went on to co-create Enantigen, which was a startup dedicated to developing some of our antivirals. Enantigen was eventually acquired by OnCore (Arbutus). Yanming went on to become our director of medicinal chemistry and helped spin out Harlingene Life Sciences and RimmSting, onsite faculty startups.

Between 2006 and 2007, the real wizard of developing medicinal chemistry as a partner to academics was Dr. Alan Reitz, who conceived of a different kind of Newco that he called Fox Chase Chemical Diversity Center. His approach was to become the trusted medical chemistry partner for academics and their spinouts from all over the country. He realized and could see with us from the Center that many, perhaps most, academic drug discoverers who were life scientists had no professional medicinal chemistry support. A professor who excelled in virology and developed a clever screen in their university lab might sit with a good-looking "hit" from a screen but have few means to progress the compound beyond its crude hit status. Alan's company would eventually partner with faculty from U. Penn, Drexel, U. Pittsburgh, and elsewhere. He had found a very valuable and needed niche. By 2022, his company had dozens of partners and clients from universities and small startups (that were also deficient in professional chemistry) from all over the world. His company has been responsible for several drugs, mostly for neurological disorders, reaching Phase II clinical trials, and has been involved as the chemistry component of multibillion-dollar acquisitions, such as Biohaven.

By 2015, Drexel had exited our biotech center, and we bought out Del Val's stake in the center and its real estate. We were now completely on our own. It was just us and a biotechnology revolution.

1.5 The Pennsylvania Biotechnology Center is Located in the Best and Worst Place

I need to pause and write something about Doylestown because I am often asked why and how a life sciences incubator can be successful here. It is a small, semirural town approximately 30 miles north of Philadelphia. It is a surprising place for a major biotechnology research and business incubator facility. Nevertheless, there it is. The Pennsylvania Biotechnology Center is on Old Easton Road in Doylestown, PA. Its neighbors, when it started, were Tim's Auto and Happy Nails (for manicures, not a hardware store). Across the street was a pool contractor's dump. Literally, a dump with piles of concrete and metal wires and whatever is ripped out of old pools or did not make it into new pools. The PABC was and is near beautiful, charming, historic Doylestown. However, is not *in* the town. It is in the remnants of Doylestown's once-industrial section, which is unattractively dotted with light manufacturers, chemical and gas suppliers, and a number of old warehouses.

Doylestown as a trading center makes claims back to the time of the Lenape Indians, but the modern wave of significant western commercial development occurred largely around the time the Northern Pennsylvania railroad extended its line from Philadelphia to the borough in 1854. That's 40 years after it became the Bucks County seat of government. Even with that, Doylestown borough is still rather small. Today, as I am writing this, its population is approximately 10,000, with another 20,000 in the surrounding township. It has no major academic or government research institutions (the local university is primarily an undergraduate institution) and certainly no major airports. The closest would be in Philadelphia and Princeton, NJ, which are more than 30 miles away. Its demographic profile is also not representative of the rest of the country (see Table 1.1). It is not much of a slick tech corridor and an exceedingly unlikely place for a life sciences business incubator. However, today, the PABC is one of the nation's most successful.

On the other hand, although Doylestown itself is not much of a biotech hub, it is essentially the bedroom community for many pharmaceutical and biotechnology companies. Thousands of Ph.D.-trained scientists and executives who work at Merck, Novartis, Roche, and J&J live and sleep in Doylestown, although they work 15-50 miles away. Possibly because of this demographic, there is solid public and regional governmental enthusiasm for research and entrepreneurship in the county.

Solid as the support from the community might have been, when we opened in 2006, the 17 of us who had been working in the rented space on the college campus sat nervously in the 60,000 ft^2 nearly empty warehouse in which we relocated. It had been converted into sophisticated new offices and labs, but it seemed to me we were very alone. Those Ph.D.s and pharma executives who may have been spending their nights sleeping in Doylestown were all at their jobs, somewhere else, during the day. We were alone. But things have significantly changed. In 2023, the PABC is now a 14.5 acre, 4 building complex with more than 150,000 ft^2 of lab, office,

Table 1.1 Demography of the home of the Pennsylvania Biotechnology Center[a]

	Doylestown	Bucks county	US
Demographics			
Population	8352	645,054	333,287,557
Gender female	52.9%		50.5%
Age			
Under 5 years	4.1%	4.7%	5.7%
Under 18 years	14.3%	20.1	22%
65 and older	27.5%	19.89	16.8%
White	92.7%	87.4	75.8%
Black or African	1.3%	4.7	13.6%
Asian	2.4%	5.5	6.1%
Hispanic	2.1%	6.1	18.9%
Education[b]			
High school or more	97%	94.5	88.9%
Bachelors or more	59.2%	42.9	33.7%
Income			
Median household	$86,188	$99,302	69,021
Per capita	$55,262	$50,315	37,638
Poverty	7.5%	6.5%	11.6%

[a]U.S. Census 2020
[b]Highest level of Education achieved for those 25 years and older

instructional, and meeting space. There are more than four hundred people working in 41 startups, along with the nonprofit HBF and its Blumberg Institute, which manages the PABC. However, in 2006, there were just 17 of us.

1.6 The Biotechnology Industrial Revolution Happened Despite Scientist Self-Consciousness

Hard to believe now, but there was once a prejudice against making money from your research at universities. This was certainly the way it was when I started my career, first as a graduate student at the State University of New York at Buffalo in the late 1960–early 1970s and then as a postdoctoral fellow at Princeton University in the late 1970s. For an academic life scientist, it was considered to be in very poor taste to even think about the commercial potential of your research, let alone commercializing the research yourself. In fact, one professor of mine at Princeton went so far as to boast that he worked on things that had "no commercial value," and *"that was the point."* He exaggerated (I hope) to make his point. Ironically, he went on to make some very important discoveries that were significant contributions to the practice of academic and industrial molecular biology. In addition, I believe he went on to make quite a bit of money.

There can be no question that in those days, at least in the life sciences, there was a sense that university research needed to be pure science, and the distinction between commercial and pure science needed to be very clear and sharp. It was vulgar and crass to blur those lines. This wasn't limited to where I was. This was pretty much the academic culture in life sciences at that time. I am careful to specify life sciences because, at the same time, I believe engineering schools were full of entrepreneurs and commercial, applied research. Life science researchers thought of themselves as noble knowledge warriors, willing to devote themselves to abstract matters of uncertain practical value, where the only motive was increasing human knowledge. As self-conscious as that sounds, there was a good deal of truth to it. It was not clear or necessary that there would be new medicines discovered by researching basic biochemistry or microbiology. The caverns of science led to uncertain destinations.

However, there was gold in those mines. And this eventually became too irresistible to ignore. In the U.S., attitudes were changing. Perhaps it was legislation such as the Congressional Bayh Dole Act of 1980. Or a change in culture such as an increase in risk investment interest from the financial community or a new generation of entrepreneurial faculty. Or perhaps it was the irrepressible, irrefutable value of some of the major discoveries (i.e., molecular cloning of medically important products). Whatever the reasons, the 1980s and 1990s became the decades of the first Biotechnology Industrial Revolution Gold Rush.

By this, I mean the time when translational research at universities thrived and led to unprecedented growth in contributions to biopharma development. There are numerous examples of this from that period. Human insulin was molecularly cloned and expressed in *E. coli* in 1978 and 1979 by Arthur Riggs, Herbert Boyer, and Keiichi Itakura. This became a recombinant therapeutic, developed and sold by Genentech and Eli Lilly in 1982. Recombinant DNA and molecularly cloned proteins became images for the covers and centerfolds of magazines (Fig. 1.1).

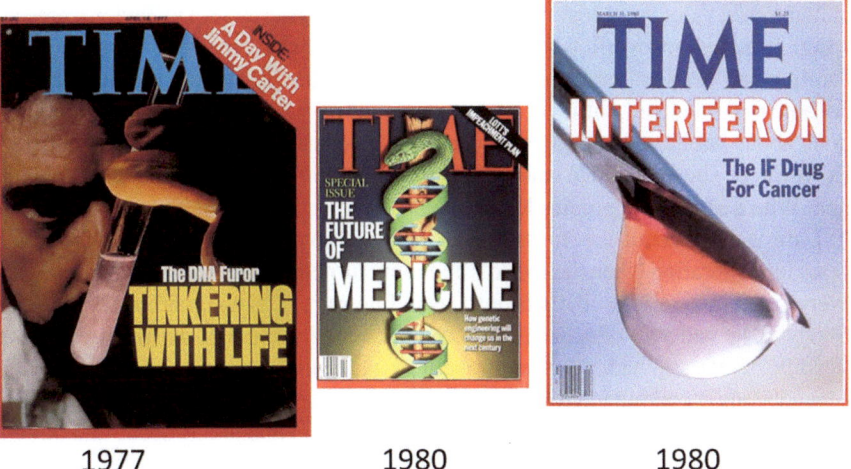

Fig. 1.1 Image of the cover of Time magazine from 1977, 1980

Recombinant DNA and interferon were covered for *Time* Magazine in 1977 and 1980; interleukin was featured on the cover of *Fortune* Magazine in 1985; insulin in the early 1980s; and recombinant DNA on *Time* in 1999.

In many respects, the 1990s and into the early 2000s were to biotech what the nineteenth century industrial age was to manufacturing. However, in 2000, I had little to no experience with or awareness of business incubators. My career at that time was as a medical school professor and research scientist. That said, I did have a great respect for and interest in applied biology and admired those who could take something from discovery to practical use. Okay, even commercialization. Therefore, unlike many of my basic science colleagues, I did have my eye on all of the biotechnology commercialization fuss going on. In addition, this all came together in place and time to the benefit of our newly created business incubator in Doylestown. It gets better. Or worse. Depending on what you were doing at that time.

If 2006 was an uncertain time and many found our newly renovated but empty Doylestown warehouse an uncertain place for a biotechnology Incubator, by late 2008 our location turned out to be a tremendous asset.

I had invited Herbert Taylor, Ph.D., an economist and one of the Philadelphia members (the governor) of the Federal Reserve Bank, to give a business seminar about the role of the Federal Reserve in economic growth, particularly as it relates to biotech. This was scheduled for December 2008. At the time of my invitation, I was worried his presentation would be too technical and uninteresting to the scientists and entrepreneurs at the Center. In addition, no one would come, or worse, they would come and leave before Dr. Taylor finished. So, I tried to create a bit of excitement to draw a crowd, but I never imagined what would happen between the date of my invitation and December 2008 that would make promoting his presentation unnecessary.

Dr. Taylor's presentation was one of the most relevant and best attended seminars we ever had. At the time of the invitation, the value of the Dow Jones Industrials was hovering at approximately 13,500–14,000. By the time of his presentation in December, the Dow had dropped to ~8,000, and we were all in the middle of the greatest recession since 1929. More than eight million jobs and almost $19 trillion in household value were lost. People were panicked. Therefore, Dr. Taylor's presentation attracted an overflow crowd as well as print and broadcast news media. I barely recall his seminar beyond appreciating that he did show up. I knew he had become overwhelmed with work, and people in his position had become very cautious about public remarks. It would not have been surprising if he had canceled. However, he showed up. He was a celebrity by that time. He gave a terrific presentation. Who knew an economics lecture could be so enthralling.

I am writing about this to point out that our biotech center was only a couple years into operation when the world's economy turned upside down. Parenthetically, every organization was charged rent, including ours, Del Val, and, of course, the Newcos. We had a lot of space to heat and service. The biotech center was growing steadily, but we still had, what we euphemistically call, excess capacity. Our success depended upon continued growth, either from our research enterprise (Drexel, HBF, and the Blumberg Institute) or from startups coming in and taking space and paying rent. With such a severe economic downturn, there was uncertainty about the viability of our current tenants, let alone whether there would be any Newcos paying rent.

During the 2008 recession, paradoxically, the Center actually thrived. First, the Drexel and Blumberg Institute scientists provided a continuous source of Newcos. In a sense, we were creating our own tenants. An economic perpetual motion machine. Quite an unusual business model. The startups that continued to be spun out by our faculty were able to obtain nondilutive federal and state grants. Angels (early investors) were also still around. These Newcos were a source of rents and additional intellectual life for the Center, and, therefore, a perpetual motion machine of ideas, company creation, and rents that feed the Center would lead to more ideas, more Newcos and new jobs. In addition, there was an enormous parade of pharma scientists and executives coming to us with ideas, connections to other scientists, and a network of funders looking to start companies on their own. Many of these individuals had been laid off, many had seen the writing on the wall, and others had been ushered into retirement early.

In hindsight, I should have been able to predict that a recession was coming. Early in 2007, there had been a growing number of outstanding, innovative scientists coming to the Center, looking to either start a Newco or somehow get involved with one. This must have been the tip of the recession spear, and had I known better, we could have anticipated something economically rotten was coming because of the increase in those who had been laid off and were looking for opportunities with us. We tried to accommodate whatever was needed. This was sometimes lab space, offices, or just desk space. We often gave gratuitously while they wrote business plans, SBIR grants, and sought funding. Many did quickly get funding for their Newcos, and others either joined the existing funded Newcos at the Center or joined us, or eventually bounced back to other big pharma. During the recession, these numbers just increased, and we actually experienced some of our best growth with bright and innovative companies being created out of necessity by many individuals who had been carrying ideas with them for years and now had a chance to give them a try.

1.7 Celebrity Science: Baruch S. Blumberg and W. Thomas London

The Pennsylvania Biotechnology Center (PABC) became populated with innovative entrepreneurs from the pharma community and academia. Many of these individuals were leaders in their fields and could provide mentorship and inspiration. Among the "celebrity" scientists was Nobel laureate Dr. Baruch Blumberg himself. We gave him an office at the Center, and for several years he would hold office hours on a monthly and sometimes weekly basis. He had retired from NASA and Fox Chase Cancer Center, and we were lucky he considered us his new professional home. He inspired the faculty and excited the students. He was always sharp, thoughtful, and had an encyclopedic knowledge of, well, it seemed everything. His unlimited knowledge was quietly demonstrated in a trivial pursuit-type question-and-answer game that Dr. Tom London led one day during a seminar. Dr. London served on the HBF Board and was also a regular presence at the Center. He was a well-known and accomplished physician-scientist who worked with Dr. Blumberg at Fox Chase Cancer Center and was credited with helping make the association of hepatitis and the Australia antigen in the 1970s.

During a special seminar about Charles Darwin for our faculty and trainees, having re-read the Origin of the Species and a book about Darwin, Dr. London was excited and eager to hold a discussion. The seminar was off-beat, obviously different from our regular series, and done in fun, with a pop quiz thrown at the audience. The audience had some pretty sharp people. I was sitting next to Dr. Blumberg, who consistently answered every question quickly and correctly before anyone else could even guess. Truly. Every question. I had to ask him to slow down so others could give an answer. He chuckled, but continued to answer every question correctly, mouthing the answer to only me. He really did seem to know it all. Or at least he seemed to know everything about the natural world.

Dr. Blumberg passed away suddenly in 2011 while giving a keynote presentation at a NASA conference in California. His death rattled me, almost as much as the passing of my father and mother. He had become a very familiar, reassuring, encouraging, and wise presence. Our Board unanimously agreed in 2013 to rename the HBFs research organization, previously called The Institute for Hepatitis and Virus Research, the Baruch S. Blumberg Institute. Dr. London passed away a few years later. This was another shock and upset. We honored him by establishing the W. Thomas London Chair, which is held by a physician-scientist with whom he worked and told us to hire: Ju-Tao Guo, M.D.

Dr. Guo, who worked at Fox Chase Cancer Center directly under Christoph Seeger, Ph.D., and with Dr. London, became the first scientist to receive a named, endowed chair at the Blumberg Institute. When we were considering recruiting Dr. Guo from Fox Chase, Dr. London told me that Ju-Tao was the smartest scientist he ever knew. He said that directly to me, overlooking a possible insult to me. But I agreed. He was indeed very smart, and the right person to hire to help lead the HBV research programs.

Dr. Guo had come to work with Drs. London and Seeger in the late 1990s from Beijing. From the mid- to late 1980's, there was a thaw in tensions between China and the U.S. with increased trade and immigration. Between 1980 and 2018, for example, the number of Chinese legally living in the U.S. increased sevenfold (Cebolla-Boado and Nuhoglu Soysal 2023; Zhang et al. 2022; Echeverria-Estrada and Batalova 2020). In addition, there was a noticeable, and I think, very beneficial, increase in Chinese students and scientists coming to the U.S. for training. American universities and research organizations, both for-profit and nonprofit, swelled up with students and scientists seeking education and opportunity from China, South Korea, India, and even Russia. Ju-Tao was part of this wave, which, to my mind, has resulted in advances in U.S. technology. Since our work focused on HBV and liver cancer, which are major public health problems in China, we tended to attract a disproportionate number of scientists from Asia. It was obvious that between 1990 and 2015, the U.S. was an appealing place for Asians to learn and work, and we were lucky for those who chose us.

Ju-Tao was everything Dr. London had promised. He started with us when we were at the Jefferson Center, on the campus of Del Val, and initially had a pretty low profile, working directly with me. However, what a mind he had, I soon discovered. When we moved to the PABC, he was provided with his own lab. He molecularly cloned one of the first productive hepatitis C replicons (modeling Bartenschlager and Rice), produced reliable and robust antiviral HBV assays that ultimately led to the identification of some of the first HBV capsid inhibitors, which became the basis of several company spinouts at the Center and is a class likely to be eventually approved for use in managing chronic hepatitis B.

1.8 The Leap: Leaving the University

It was in 2015 that I took a final big professional leap, in which I formally left the university for our nonprofits at the Pennsylvania Biotechnology Center. After that jump, for the first time in my professional life, I was no longer a university employee. Although in my career there were a number of steps taken along the way that were unconventional, for me this was the one that was absolutely identity altering. It was my great leap. In 1997, I moved my lab from a traditional setting at TJU to our rented space on the campus of Del Val to establish the Jefferson Center for Translational Research. That was one step, but I was still a professor at TJU. Then, in 2005, while still on the college campus, as explained earlier, I moved my full-time faculty appointment from TJU to Drexel University. I was still a university professor. It was not until 2015 that we completely separated from Drexel and all traditional university affiliations, and we joined the HBF and the Blumberg Institute as employees. The safety net of tenure at a university was completely gone. The reason for this was complex and opportunistic.

Drexel had initially been fully supportive of our physical relocation from Del Val to the soon-to-open PABC, located about 3 miles from the small college. In addition, we thrived. Although we were extremely productive, by the beginning of the second decade of the 2000s, the climate at Drexel had changed. Dr. Papadakis, who recruited me, passed away in 2009. By 2012, Dr. Bill Stephenson, who hired me, had also passed away, and the Dean who worked with me (Richard Homan, M.D.) was gone. The Drexel Medical School had a new Dean and although our relationship was cordial, and my group continued to be one of the most productive in the school, just as we experienced with TJU, Drexel wanted to bring us all back to the city campus. There had been a number of departures of both faculty and decreases in overall grant funding at Drexel, so perhaps paying for a satellite program such as ours made us less appealing, even if we did more than cover our costs.

To be fair, Drexel would have been justified in wanting to consolidate and bring its faculty together. However, I think the new university administration was also nervous about all of our entrepreneurial activities, conflicts of interest, and potential fundraising loyalties being divided between the PABC nonprofits and Drexel. The Blumberg Institute was growing, as was the PABC. This may have also been unsettling to Drexel leadership. Frankly, I was not unsympathetic. The arrangement between Drexel, HBF, and the Blumberg Institute, the PABC, and the growing number of startups was unusual for academic institutions. The fact that we were 30 miles from their main campus and that the head of their offsite academic division was also the head of several nonprofit organizations made things all the more uncomfortable for them. While I was confident about our virtue and we brought in significant revenue and scholarship with $4–5 million a year in excess funding and 20–30 publications per year bearing the Drexel name, it wasn't difficult to see how new leadership would prefer to have all of this activity back on their campus, where there was excess capacity. Presumably, we were kind of like cowboys to them. Being on the frontier, on our own, made them nervous. If you knew me, if you knew us, you would know how wrong that image is.

The initial proposals from Drexel had the university acquiring the HBF and the Blumberg Institute, though allowing them to continue to operate in Doylestown for the foreseeable future. Since the foundation was understandably uncomfortable with that, another idea was for me and the other Drexel faculty onsite to relocate to labs on their Philadelphia campus. Although Drexel offered to keep our division designation intact, the HBF and its labs would remain in Doylestown at the PABC. This would have been very damaging to the Blumberg Institute, since our programs represented a significant amount of the intellectual and financial life of the Center.

Ultimately, we decided this was an opportunity for the HBF. It was a chance to make a break. If the funding authorities agreed and my research colleagues would join me, I could transfer my programs to the Blumberg Institute. Although I was no longer a principal investigator for 100% of the research funding for our programs, I was still close to 75%. This would mean the Blumberg Institute would instantly have a $4–5 million a year budget and 20–30 employees, including faculty. However, it would need to assume all of the responsibilities of an independent research institution, from all the myriad compliances to purchasing. That would be a challenge given our size and remoteness from other research organizations.

The timing was as good as it was going to get. The company OnCore (now Arbutus) was being created, in large part with Drexel and the Blumberg Institute's technology (i.e., our inventions), and a generous sponsored research agreement was being negotiated. OnCore was flush with cash from its lead investor, Vivek Ramaswamy, who had made a fortune with the biotech company, Pharmasset. Mr. Ramaswamy was brilliant, brash, and, in 2015, not more than 29 years old. In 2023, he became phenomenally and suddenly famous as a surprise candidate for the 2024 presidential election. Back then, he seemed committed to HBV research and was himself in awe of OnCore's chief scientist, Dr. Mike Sofia, who was the inventor of sofosbuvir, Pharmasset's HCV antiviral cure. Dr. Sofia's invention was the basis for Pharmasset's $11.5 billion acquisition by Gilead Sciences, which made Mr. Ramaswamy very wealthy. OnCore and Mr. Ramaswamy wanted to continue to support our work (particularly the work of my colleagues, Drs. Ju-Tao Guo, Jinhong

1.8 The Leap: Leaving the University

Chang, and Yanming Du). Moreover, the company we created, Enantigen, was being acquired by OnCore. Therefore, there would also be new injections of cash from the sale of that company. The Blumberg Institute could now really thrive and expand.

Although this was a rare opportunity for the Blumberg Institute, it was a bit risky for me. I had a "celebrity" tenured chair at Drexel, which was the old-fashioned tenure, in which my salary was guaranteed for, well, my life. That kind of tenure is now very rare to find, and although I don't want to exaggerate the sacrifice, my tenure had significant value. Tenure of any kind has become controversial, and no surprise, I was and am one of its defenders. Those who took positions with tenure generally expected less compensation than they would have received otherwise. But the purpose of tenure is to embolden its holders to pursue work that might be unpopular or otherwise unsupported. I like to think I was so motivated, to the benefit of TJU and then Drexel. The bottom line is that I valued my tenure and proved the value of tenure to my schools. Moreover, my lifelong professional identity was wrapped up in being a professor and in an academic culture. Giving it up for an independent research institute, even one I created (perhaps especially one I created), with an uncertain future, made me pause, a bit. My wife, Joan, was all in.

Fortunately, most of my onsite Drexel colleagues working on antivirals were in. That is something about which I, the HBF, and Blumberg Institute should be appreciative. In a show of confidence, they took the leap, too. Drs. Ju-Tao Guo, Jinhong Chang, Ying-Hsiu Su, and Yanming Du all joined me in resigning from their positions at Drexel and accepting positions at the Blumberg Institute. The separation from Drexel did create a diaspora. Dr. Haitao Guo took a faculty position at the University of Indiana, and Drs. Anand Mehta, Pam Norton, and Pooja Jain remained with Drexel, moving 30 miles to the downtown campus.

Dr. Norton focused on coordinating graduate programs at Drexel, continuing much of what she had started with us in Doylestown. We had a robust Master's program that attracted suburban, part-time students with pharma jobs. It struggled during the recession because pharma stopped paying for whatever staff remained to take graduate courses. This program had never fully recovered for us, and was something I wanted to address, so losing Dr. Norton complicated that effort. Dr. Norton was an intelligent and rational voice of reason at our research meetings. She moved to Drexel's campus, ultimately helping build an online Master's program that was a very significant part of their academic revenue for the Microbiology Department.

Dr. Mehta, my closest scientific partner, who had been with me on and off since immediately after his undergraduate days, felt it was time for independence. At Drexel's urban campus, he was being treated as a new faculty hire, which meant a significant investment in getting him set up. Anand received a generous startup package, including lab renovations and new equipment. Most importantly, he received a tenure track Associate Professor title with the promise of tenure review in a few years. It appeared to be a very good deal for him. He took it, but we continued to collaborate.

Major research universities typically provide starting packages to the faculty they are recruiting, and these packages can reach multimillions of dollars. The package may include renovation of lab space, new equipment, and salaries for the professor and staff for several years. In 2015, for example, a lab renovation could easily cost $500,000-1 million. New equipment could add another $500,000. Three years of lab supplies and staff support add another $500,000-1 million and a guarantee of the professor's compensation at $150,000–200,000 per year. I was offered a $3–5 million package to move to a major southern university in 2006 but chose to remain at Drexel to build the PABC. The offering university assumes that the professor will obtain or bring with him/her millions of dollars in extramural grant support, which would cover much of the professor's compensation. Even if the professor is well funded, there is usually a net loss to the school, but of course, the professor's research and teaching are fulfilling the university's mission. In addition, a celebrity professor will attract other faculty, students, and grants, and even boost a clinical reputation. By the way, when I moved to Drexel, there was absolutely no net cost to Drexel.

Anand had been working with me for decades, first when he was right out of college and then with me at Oxford, where he ultimately completed his doctorate. I did much of my thinking with and through him, and his leaving was like a matrimonial separation. As with all trainees, there comes a time when most want to leave and develop their own identity, not that mine was that overshadowing. Therefore, Anand moved his lab downtown to the Drexel campus. But not for long. Within two years, he was scooped up by the Medical University of South Carolina (MUSC), accepting an endowed, tenured chair in proteomics. The MUSC offer was perfect for him. I was an admirer of MUSC and had been eyeing that school in charming Charleston, SC, for two decades. I even had discussed a fantasy with the Vice Dean in which we would have a research group that would spend summers in Doylestown and winters in Charleston. He thought I was kidding. I wasn't. With Anand there, I felt that part of me had found its way. Anand would continue to study liver cancer glycomics and establish one of the leading liver cancer glycobiology labs in the nation.

Dr. Haitao Guo was another extraordinary talent and an important part of the HBV antiviral team. The offer from Indiana University was too good to refuse and a chance for Haitao to establish an undeniably independent lab. However, interestingly, similar to Anand, he did not stay long at the first post-Drexel job he took. Within a few years, he was back in Pennsylvania as a full professor at the University of Pittsburgh, where he is now a well-funded leading HBV researcher.

1.9 The Golden Age of HBV Drug Discovery

Although our faculty numbers were diminished by the departure of several Drexel scientists, our new funding, mostly from our windfall support from our biotech enterprises (i.e., OnCore/Arbutus and the acquisition of Enantigen), allowed us to hire several other researchers. There is still a very formidable concentration of

1.9 The Golden Age of HBV Drug Discovery

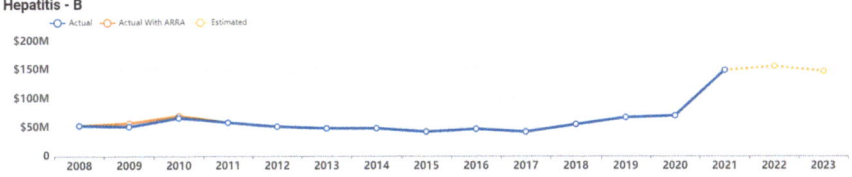

Fig. 1.2 NIH Funding for HBV (in millions of $). (Source: NIH NIAD and NIH online data)

hepatitis B scientists. The timing for this concentration turned out to be fortuitous. In 2015, interest in HBV was suddenly on the rise.

I would like the Hepatitis B Foundation to take credit for this increase in interest and to some extent it can. HBV research has always taken a back seat to other diseases. Research funding for HBV from the NIH in 2015, for example, was slightly more than 1% of that spent on HIV, and it was also dwarfed by HCV, influenza, and certainly biodefense funding. The reasons for this are complicated. This was partly due to the assumption that with the vaccine so effective, as alluded to by the public health officials who met in Congressman Jim Greenwood's office in 1990, the problem was essentially solved. In addition, HBV is much more of a problem in Africa and Asia and much less so in the western and developed world. HCV and HIV dominated the news and the political scene. Funding for HBV was simply never prioritized.

It was the HBF's job to publicly advocate for increased HBV funding in Washington, D.C. However, we were pretty much the only voice doing this for hepatitis B. I recall speaking at a U.S. House of Representatives Appropriations committee. There were numerous HIV and HCV groups, but Joan and I were the only ones there for HBV.

We stuck with it. The 2023 levels of support for HBV research at the NIH were double what they were 5 years prior (Fig. 1.2). The addition of HBV and liver disease research to the U.S. Department of Defense Research portfolio is almost entirely because of HBF's advocacy and, of course, the sympathetic ears of several members of the U.S, House and Senate. With the help of public health lobbyists Mike Hall and Alyson Lewis of Madison Associates, we became frequent faces in the U.S. Capitol. We also regularly met with leadership at the NIH and CDC, usually around budget appropriation times. We prevailed upon our elected PA representatives in the U.S. House and Senate and other sympathetic members to ask questions of NIH leaders about HBV and liver cancer. We then helped in writing report language into the federal budget for NIH, CDC, and Defense appropriations legislation that encouraged more research programs in HBV and liver cancer.

Eventually, this report language turned into new funding and new calls for HBV and liver cancer research at the NIH, CDC, and even the Department of Defense research programs. The plus-ups were relatively small, perhaps a few million dollars a year to as much as $10 million. Compared to HIV, influenza, or biodefense it was a pittance. However, for HBV, it represented 50% increases in what was provided previously, and this had a very significant impact on the field.

Even with increased funding, NIH support for HBV was modest compared to what industry could put in. Almost suddenly, it seemed that after decades of indifference, there was a wave of interest from pharma. I mentioned the frosty encounters of the 1990s, and this continued into the 2000s. However, by 2015, things had changed. The wealth of China and growing political and economic interest in Asia overall made investments in a drug that had importance to the Asian community in the U.S., Europe, and particularly China, very attractive. Then there came a tipping point.

HCV was cured. Perhaps one of the most remarkable and important achievements in modern medicine was the discovery and development of a rather simple, safe, and orally available cure for HCV. Today, many people, even professionals, underappreciate how big a deal this was, saying that HCV was an easy virus to cure. That is rewriting history. HCV was a frightening threat to the world's public health for two decades, beginning with its discovery in the 1980s and ending with its cure in 2015. In the U.S., HCV displaced HBV as the major etiology of liver cancer. It caused chronic infections, and there was no vaccine. There is still no vaccine. There had never been a cure for a chronic viral infection, and the general thinking in the virology community was that chronic viral infections were likely to be incurable. Herpes simplex, cytomegalovirus, Epstein-Barr virus, Kaposi sarcoma herpesvirus, HBV, and HCV all establish intractable, irreversible nests of infection within their hosts. The assumption has been that it would be very difficult to eradicate them without causing catastrophic damage to the host. The host is the person infected.

Interferons alone and in combination with other drugs showed some promise in slowing HCV disease. There were even examples of people who were cured or experienced sustained suppression of the virus and disease. However, this was uncommon. Similar to HBV, the systems for discovering antiviral agents are very limited. The breakthrough came with the introduction of the HCV replicon from the labs of Dr. Ralf Bartenschlager at Heidelberg University in Heidelberg, Germany, and Dr. Charles Rice at Washington University in St. Louis, Missouri. Not a true infection model, but as Dr. Rice had developed for other RNA viruses, a very clever system in which the viral genome could self-replicate inside a cell in tissue culture. This allowed for high-throughput screening of compounds that prevent HCV replication. Drug companies went wild with this.

There were several new drugs introduced clinically that inhibited HCV, and the percentage of people with HCV that could be cured or at least achieve sustained virological suppression increased. It was Dr. Mike Sofia at Pharmasset and his team's discovery of sofosbuvir, which inhibited the HCV polymerase, that most dramatically changed the clinical picture and the way we think about the possibility of a direct-acting antiviral agent's ability to cure people of HCV. Other drugs, such as daclatasvir, from the lab of Dr. Min Gao at Bristol Myers Squibb, were also remarkable and even had a very novel mechanism of action by interrupting HCV protein organization. These other drugs also made it clear that HCV cures are not only possible, but when used in the correct combinations, cures are almost certain.

After as few as 6 weeks, but never more than 12 weeks, sofosbuvir or daclatasvir taken together can be curative 98% of the time. HCV is no longer a frightening,

1.9 The Golden Age of HBV Drug Discovery

incurable disease. Some now diminish the significance of this cure biologically, saying an RNA virus would of course be easy. However, that's not true. The fact that a direct antiviral can cure a chronic liver disease is a paradigm change, both medically and biologically. In 2016, Drs. Bartenschlager, Rice, and Sofia won the Lasker Prize for their work. In 2020, Drs. Rice, Michael Houghton, and Harvey Alter won the Nobel Prize in Medicine for the discovery of HCV.

With a cure for HCV, people began to think a cure for HBV was possible. Moreover, there was an exodus of HCV experts to, well, HBV. However, not so fast. While HBV and HCV chronic infections share clinical similarities, as explained, there was little similarity at the virological level between HBV and HCV and even less at the epidemiological level. People leaving HCV research for HBV were not entirely logical. Regardless, that's what happened, and by 2015, with HCV curable, there was a tsunami of commercial interest and, hence, investment in HBV.

The HBF and the Blumberg Institute were national centers of HBV excellence. We were authorities on the subject and had a war chest of anti-HBV assets, including some of the only, if not the best, antiviral assays for HBV. Most importantly, we developed assays that could target steps in the viral life cycle that had never been previously targeted, such as HBV cccDNA, HBV capsid, and HBV surface antigen (HBsAg).

Our technological advantages with HBV made us attractive to investors looking to create new companies targeting HBV, as well as a physical location for startups with an interest in virology. It was an exciting, wild time for us, where it seemed an entire biotechnology ecosystem of creation, growth, merger, and acquisition was happening. Enantigen, a company created with our technologies, had a parade of commercial suitors. OnCore had also been created with our technologies and had the cash to acquire Enantigen. Novira, a company created by Merck scientists and led by Osvaldo Flores, Ph.D., to develop HBV drugs, was also at our center and received a tremendous, multi-million dollar investment. By 2017, Novira would be acquired by Janssen Pharmaceuticals, a J&J company, for a rumored ~$670 million.

And it wasn't just HBV-oriented companies that were taking off at the Center. Synergy Pharmaceuticals, which Drs. Blumberg, Dwek, and I created to pursue HBV, eventually obtained FDA approval for its drug. This was a major milestone for a startup biotech company. The Synergy drug was created at our Center by Dr. Kunwar Shailubhai and his colleagues Dr. Gary Jacob, Synergy CEO and scientist. Under Dr. Jacob's leadership, Synergy took this drug from invention through all the preclinical steps and then through the jaws of clinical trials. It is difficult to express just how remarkable this is. There are usually only 30–40 drugs approved by the FDA each year. These are achievements that very few companies other than those of big pharma have been able to accomplish. From the universe of smaller biotech companies, I can think of perhaps only 10 that have been able to discover a drug and take it all the way through approval. Synergy, at our Center, was one. That said, although Synergy was a proud success in many ways as a discovery and development phase company, it eventually grew to more than 400 employees with a robust sales force.

The year 2015 can certainly be considered the middle of an HBV Golden Age. The HBF was gaining national attention for its public health and advocacy efforts, led by my wife Joan and now her successor Chari Cohen, DrPh, MPH. For example, getting HBV listed as a protected condition under the Americans with Disability Act (ADA) was a significant victory, both symbolically and practically. The initiative started with Joan after learning several medical students in U.S. schools had been asked to leave their programs because of their HBV status. In addition to this being heartbreaking and career destroying for them, it was outrageous and unnecessary, both medically and from a public health perspective. In 2011, Joan and Chari took up the cause, and the HBF worked with the CDC to update its 1991 recommendations for HBV-infected healthcare students and workers based on the effectiveness of existing antiviral treatments to significantly reduce the risk of transmission. The CDC guidelines were updated in 2012 and became the cornerstone of the U.S. Department of Justice's (DOJ) lawsuit against one of the medical schools that rescinded their offers to two students who had chronic hepatitis B. The DOJ won the case in 2013, and a legal precedent was established. The ADA protection against discrimination now includes people with chronic hepatitis B. The medical students were readmitted and finished their education. New federal policies and laws were established as well. Healthcare students and workers can no longer be rejected or removed because of chronic hepatitis B. I mention this, here, as an example of both the good work of the HBF and its growing national impact.

HBF, Blumberg Institute, and the PABC were having their own Golden Age and a bit of new wealth themselves. This allowed me to leave Drexel and move from being a volunteer president and CEO to a paid employee. We were also able to recruit new business professionals to run our organizations, such as our first COO, Jim Horan, and then Nelson Carvalho. Now we were no longer depending on me to figure out how to keep the Center's lights on, water running, and collect rent.

No question, our enterprise was experiencing a growth spurt. The labs were busy, the halls were bustling with people, and the conference rooms and seminars were plentiful, well-attended, and lively. It was thrilling to me to have people attending our seminars such as Drs. Michael Sofia, discoverer of sofosbuvir; Patrick Lam, father of the Lam Reaction and lead discoverer of Apixaban; Bruce Maryanoff, discoverer of Topamax; Min Gao, discoverer of daclatasvir; and Kunwar Shailubhai, discoverer of Trulance. Also seen in our halls were Drs. Thomas Shenk, member of the National Academy of Sciences and Princeton Professor who, with his wife Lillian Chiang, started the company Evrys at the Center; Dr. Osvaldo Flores, founder of Novira, which was acquired by J&J; and many other outstanding scientists, often all in the same room (not to mention our mighty research team). It's not an exaggeration to say that we had a high concentration of VIP drug discovery researchers that was probably unrivaled in any nonprofit, small incubator setting.

Around 2013, the Blumberg Institute had another remarkable acquisition. I had a chance meeting with my neighbor Dan Ring, a Merck executive, that led to the donation of their entire U.S. natural products collection. This was kicked off with a visit from the Executive VP (and old friend from TJU), Roger Pomerantz, M.D., and their storied former Merck CEO, Roy Vagelos, M.D. Briefly, Merck had lost interest

1.9 The Golden Age of HBV Drug Discovery

in natural products as a primary source of new drugs. Natural products are metabolites from living cells, in this case, bacteria and fungi. Even at this time, perhaps half the drugs that enter clinical use have their origins in natural products. Merck's lipid lowering drug, Zocor (Atorvastatin), is an example, and there are hundreds of antibiotics and anticancer drugs derived from natural products that have reached human use. However, natural products can be difficult to synthesize or rationally improve, are often of relatively low potency, and have fallen into disfavor by many, actually most, pharmaceutical firms. As a basic sciences research organization, we believed the Merck Collection could offer a new avenues of discovery, and given the collection's size and the quality of its curation would instantly leapfrog us into world leadership in natural products.

Attorney Joel Rosen was the HBF Board Chair and led negotiations with Merck. Merck was cooperative and saw this as a way to both lighten their load (maintaining the collection and remaining staff was costly) and achieve a public relations win. Merck agreed to give us the collection plus approximately $1 million to set up operations since we were a nonprofit organization. We hired several Merck scientists who had been working on the collection and built a 2000 ft^2, minus 20-degree walk-in freezer room to accommodate the collection. We became the new owner of one of the largest, most biodiverse, and best-curated natural product collections in the world.

We screened the natural products collection extensively for compounds that had anti-HBV activity and other activities. Unfortunately, we did not find any useful leads. Unraveling the actual active compounds in a natural product broth was beyond us, except for drugs with anti-Dengue virus activity. That drug was actually already in the literature, so we achieved only a publication victory. On the other hand, we set up a service in which others (usually companies) would pay us for access to the collection. This never fully took off, although there was a steady stream of business. The Blumberg Institute scientists working on natural products were excellent, and they enriched and expanded our talent pool to disciplines we would not otherwise have explored. However, the fee-for-service model they were using was not all that contributory to the organization's financial health. They were isolated from the rest of the Blumberg Institute and ran significant budget losses every year. It became difficult to justify, and we sought alternatives.

In 2014, Nelson Carvalho announced he wanted to retire from his position as our COO. He did a fine job of stabilizing operations, seeing us through the separation from Drexel and Delaware Valley College. Using a search firm that specialized in finding leadership for nonprofits, we were able to recruit Lou Kassa as our new COO. Lou had a social services background, managing agencies that worked with and housed troubled youth. He was the businessman for the youth agency but truly had no experience with academics, biotech, or disease mission-driven nonprofits. Perfect. Ironically, his background with troubled youth and people in conflict turned out to be ideal.

Lou came to us full of vigor and with fresh ideas. He was extremely respectful of the scientists working in our natural products unit and recognized the value of the collection. However, he also, frankly, could be a bit more dispassionate about it than was I and saw it as an asset that could be sold. In addition, within a year, we had several buyers of the collection. All were willing to keep the collection in operation. We raised enough money from the sale to keep the remaining natural products chemists employed at the Blumberg Institute. Ironically, much of the collection remains with us, housed in our walk-in freezers, which are now subsidized by the purchaser who also uses our natural products chemists and microbiologists through reverse contracts.

1.10 Initial Public Offerings on Research Way Boulevard

By 2017, the PABC was booming. There were hundreds of millions of dollars raised by companies at the Center through venture, partnerships and initial public offerings (IPOs). Usually, a Newco created at our Center raises funds from individuals or venture funds. The public in general has no opportunity to invest, and the owners of the company cannot sell their stake easily. An IPO is the mechanism whereby a company can sell shares of itself (equity stakes) to the public through public capital markets, such as NASDAQ. Although there are certainly other ways companies can raise funds and investors can benefit financially, an IPO is an unambiguous milestone. It's usually the long-awaited exit in which early investors can sell their stake.

It was satisfying that several Newcos at the PABC that had IPOs had direct relationships with us at the Blumberg Institute. For example, by the time of Synergy Pharma's IPO, we (meaning Drexel, HBF, and me) had no stake in the company. Their merger with a company already listed on the NASDAQ allowed them to raise scores and perhaps hundreds of millions of dollars. This is still a tremendous success story for us, but it meant little financially for us. The structure of the deal ended up causing our holdings to be almost valueless. Parenthetically, Synergy executives did feel bad about how I and the nonprofits lost out and offered me some equity, but I declined in solidarity with my co-founders, Drs. Blumberg and Dwek, and our respective institutions (I know, this sounds unbelievable). The investors ultimately did grant the HBF some shares presumably to honor my role as a company creator, and although this wasn't much, it was a generous gift by our standards, and I count it as a silver lining.

Synergy has not been the only public offering at PABC. In 2015, OnCore merged with Tekmira, which was an siRNA company in Vancouver, Canada, that shared a common interest in HBV. Funny in a way, because, as mentioned, some of the very first siRNAs came from our Center in the late 1990s and early 2000s. The merged company was named Arbutus. Tekmira's siRNA asset (which was deemed a dark horse for HBV at that time) and the much more valued OnCore assets (from us) were introduced at the JP Morgan Conference in San Francisco, and about $350 million was raised through its listing on NASDAQ. Arbutus, operating from our Center, was instantly an extremely well-funded research organization. They hired outstanding scientists, including some of the most accomplished medicinal chemists and virologists in the business, bringing them to our campus. They were nicely

1.10 Initial Public Offerings on Research Way Boulevard

Fig. 1.3 Biotech success stories from the Pennsylvania Biotechnology Center

integrated into the Center's knowledge community, enriching the environment, just as I had always hoped.

This cycle of company creation, fundraising, and the recruitment of excellent scientists and state-of-the-art equipment repeated itself many times at the Center and was a source of constant excitement. Synergy, Arbutus were early examples, but Novira, Callidus, Antagene, IGM, Evrys, and others were also impactful (Fig. 1.3). Hundreds of millions of dollars had been raised, and billions of dollars in value had been created. The relatively short period of time between arriving at or being created at the Center and successful capital raises and exits was due to our culture and philosophy. In a 2017 study we commissioned by Centri and Cambridge Associates, it was found that the time from entry at our Center to exit with return to investors was almost twice as fast as the mean number of years at other incubators.

Our approach toward Newco creation is discussed more fully in Part II of this book, but it can be briefly summarized by saying we seek business development partnerships early. Typically, we identify a promising technology, move it quickly to a "for-profit" entity, and then spin out a small startup company run by business professionals. They can bring commercial discipline and funding, working with experts, so that the asset gains in value, will be more attractive, and can be used to negotiate with larger pharma firms.

The approach appears to be working. By 2020, we were fully occupied, with 50+ organizations comprised of 46 Newcos, the nonprofit core organizations (HBF and the Blumberg Institute), and more than 350 scientists, trainees, staff, and public health professionals on our campus. Key achievements from the PABC include three FDA-approved drugs and medical devices, and six companies have been listed on public markets, such as NASDAQ. We have become a major source of innovation, job creation, value creation, and education in the state and are consistently listed in the top ten business incubators in the nation by Kolbtree Surveys (2021) and the International Business Incubator Association (2019).

We even host a high school on our campus. Briefly, from the beginning, even when on the Del Val campus, we hosted graduate and undergraduate college students all year and had a high school internship program through the summer. These programs had grown considerably, and we regularly had more than 30–40 trainees in each of them, including a Master's in Science program. In 2018, through the initiative of our builder Tim Kelly of Norwood and COO Lou Kassa, we launched an unusual collaboration with the local high schools in which science students could take their Advanced Placement (AP) chemistry course on our campus and spend half of every day in our building and labs. There was (and still is) a full-time high school science teacher assigned to this program whose base of operation was our campus, and we designated a full-time Blumberg Institute faculty member to lead it along with our other training programs. With 30–40 high school juniors and seniors in classrooms and labs and the visuals of school buses pulling up every day, we were definitely looking like a real academic campus.

Our end of town has now become a hub of science, discovery, public health, and, inspiration. The transformation from a deteriorating warehouse to a parking lot filled with cars, and a campus and buildings thriving with people that could be seen from the street was a visible reminder to me, and I believe everyone who passed by, of how much good and how far we had taken this place. In 2023, the township renamed our stretch of Old Easton Road, the Research Way Boulevard.

1.11 A Good Time to Be a Virologist But Not So Good for the World: COVID-19

The world, and hence our Center, was thrown into panic and chaos in 2020. In January and February 2020, I recall watching the response to a new strain of SARS virus in Italy and China on TV. Hospital emergency rooms in those countries were filled with cases of people in respiratory distress. They were often put on external ventilation. Health systems were being strained, and bodies on cots were literally being parked in hallways. Anxiety was everywhere. In the U.S., we watched with a degree of horror and lots of sympathy. However, in Doylestown, I don't recall being very worried about an impact here.

1.11 A Good Time to Be a Virologist But Not So Good for the World: COVID-19

And then, almost suddenly, it seemed by March 2020, our world would be upended. The virus was now spreading through the U.S., coming in from the East and West coasts. The scenes from Italy and China were being repeated here. The news images of refrigerated tractor trailers serving as mobile morgues parked outside hospitals to accommodate the number of deaths that exceeded the capacities of the hospitals are unforgettable. The virus SARS-CoV-2, the cause of COVID-19, would eventually be managed, but not until more than one million Americans and more than six million people worldwide were killed, over a period of only 3 years, and the health and political systems severely tested.

The pandemic, its cause, and its reality became subjects of ridiculous debate. Although we certainly should be able to discuss whether the responses were correct or if they were the best, no rational thinker should, at this point, be arguing about the cause, severity, epidemiology, mortality of the virus, efficacy, and importance of the vaccines.

The development of effective vaccines to prevent hospitalization and death from COVID-19 should be regarded as another great medical and scientific triumph. The speed of progress from identification of the virus in late December 2019 and early January 2020, to sequencing the virus genome in January 2020, to the beginning of vaccine trials before the end of March 2020, and to, ultimately, the demonstration of efficacy and first use of a vaccine in December 2021, certainly represents one of the most rapid, effective, and productive efforts to address a medical problem the world has ever seen. The now famous mRNA vaccines from Moderna and Pfizer-BioNTech were shown to be remarkably effective (achieving protection in >90% of those inoculated). But it was the rapid mobilization of science in the world, and particularly in the U.S., with its investment of billions of dollars in vaccine development. In addition, research and public health systems across the nation and around the world pivoted from what they were doing to work exclusively on COVID-19 to end the global pandemic.

The PABC played a part, too. Several companies at the Center developed or redirected programs to detect (i.e., diagnose) COVID-19. Others began to develop vaccines and therapeutics. Of course, there was suddenly tremendous financial incentive to do this, with vast sums of federal and state funds, and even commercial investor dollars being offered. This created a world of contradictions. In the midst of a disturbing, chilling, and dystopian pandemic, there was also a flood of investment in biomedicine. The PABC was flourishing, despite the streets around us being virtually empty. The only people who dared to come out in public were mask covered, looking like creatures from the religious sects in the Middle Ages. If you happened to be out walking, as I did every day of the pandemic to work at the Center, and encountered another person on the sidewalk, there would be an awkward avoidance of each other. A cough from someone would clear a parking lot. The PABC was permitted to remain in operation since health research was still allowed to function. But we were a quiet, creepy, deserted place with very few live interactions as people went directly to and then stayed in their labs. On the other hand, there was a collective feeling of purpose, drive and appreciation that we could function. Part of me felt this sense of collective drive is what I had been wanting for the pursuit of an HBV cure. I know, it seems insensitive to think of analogizing a pandemic as a model of how people can be mobilized, but the urgency and need driving a goal is a human trait. Survival is an evolutionary mandate.

In any event, the Center was well integrated into the county and state responses to the crisis. By April, one of the Center's companies, with a CLIA lab designation (i.e., a type of regulated lab permitted to conduct commercial diagnostic tests), had developed a validated assay for the detection of COVID-19 exposure. This was the first such assay in our county, and a tent was set up in the parking lot where people would drive through, get blood, drawn and tested for COVID-19 exposure. The company, Flowmetric Diagnostics, took its assay on the road and was the primary COVID-19 testing service for several months in our local hospital systems, first responders, and the State University. By the end of summer, however, there were competing assays from numerous other commercial services, and the Flowmetric exposure assays became much less necessary and much less used. They evolved, however, to produce and offer specialty tests that could determine not only if a person was currently infected, but if they had actually been exposed to the virus (i.e., detectable antibodies to COVID-19 proteins that are not present in the vaccine). Again, not inventions from me or the Blumberg Institute, but a great asset for the community coming from our Center.

The HBF also pulled its weight during the pandemic. I was a part of a State Preparedness Team with me as a virologist and HBF president Dr. Chari Cohen as a leading public health expert and constant advisor to county elected officials. When the first PCR test kits and vaccines for COVID-19 were available, our campus had one of the few industry-sized low-temperature freezers, and we were able to offer storage capacity to the county.

Our public health credibility and contributions were high. However, what happened to the macroeconomy was a concern. Jobs were being lost, and there was overall economic uncertainty. Unemployment was up to 16% in Pennsylvania by April 2020 it was 20% in areas nearby. Those are depression era numbers. By that I mean the Great Depression of 1929. Therefore, it was reasonable, if slightly self-centered, to worry about whether the startups in our Center would be able to pay their rent and survive.

Fortunately most did. Some struggled, but many received grants to develop COVID-19-related programs. Those that succeeded worked with other companies at the Center that had complementary expertise. Scientists laid off from conventional companies in our area, again, migrated to us and started new companies. Then, came a wave of new money into biotech, which was seen as critical to solving the problem of the pandemic, and in 2021, we even had a new drug approval (Antengene's cancer drug Selinexor), enormous multimillion dollar raises (IGM Sciences), and hundred-million-dollar acquisitions (OrthogenRx). It was truly an embarrassment of riches and another testament to our resiliency.

1.12 Growing During the COVID-19 Pandemic

In 2021, in what is assumed to be the tail end of the COVID-19 pandemic, we opened a newly constructed, 38,000 ft^2 lab, office and instructional facility. My thinking was that this new building would allow us to consolidate all of the Blumberg Institute labs and HBF offices into one location on the campus. We were scattered throughout the facility. The new building was also to provide a large multiuser lab designed to be an "accelerator" for the truly small 1- or 2-person startups who were most dependent on core services. Large event rooms and a cafeteria were also in the plan. It was initially going to to be built at a different site on the campus and three stories (not two) and nearly 50,000 ft^2, not 38,000 ft^2. In addition, the initial cost estimate was closer to $10 million, not the more than double that it eventually cost. It ended up being different in many ways. First, conflicts with our one-time partner, Del Val delayed construction. These delays as well as our underestimations of the amount of ground site preparation and material and labor costs, resulted in a much more expensive building than we had planned. Our small organization's finances were stretched to the limit.

The idea to build started with a grant I wrote for the U.S. Economic Development Authority (EDA) with a great deal of help from Konrad Kroszner, PABC Director of Systems and Engineering, and Steve Cohen, the original architect who designed the renovations of the warehouse we moved into in 2006. We initially used site plans and township pre-approvals left over from the days when the campus was a warehouse factory since we reasoned this would expedite things and make our EDA application appear more advanced and more ready. Converting the warehouse into a modern lab facility wasn't easy, and we frequently receive compliments about the conversion. Steve's experience with the warehouse conversion made him ideal for designing the new building.

The building, as anticipated in the application, was to be about $10 million, with three floors and 48,000 ft^2. The EDA would fund up to 50% of a project, so we requested the full 50%, or $4.75 million. Remarkably, we were awarded the grant in full. Nevertheless, a bank loan, as anticipated in the application, would be needed throughout the construction and for the balance. At that time I wrote the original 2001 proposal for the PABC to be on the Del Val campus because HBF and Del Val were 50-50 partners in the PABC. Although the college signed off on the grant application, their new board chair and new president suddenly refused to agree to the project. Despite there being no direct financial liability to either HBF or Del Val for the borrowing (i.e., the bank was allowing PABC to shoulder all liability), the PABC bylaws required that both partners agree to any borrowing. Therefore, the college could block the project. They had their reasons, which as far as I could tell revolved around trying to take control of the PABC.

A legal battle ensued, which was painful and public. We ultimately came to an agreement in which the college relinquished any role or rights to the PABC and received compensation for their investment. To be fair, the PABC goals and culture were not aligned with those of the college, and it is not difficult to see why they could be uncomfortable with more and more expansion. Frankly, each of our other academic partners eventually also sought to control us, so a scuffle with Del Val was just one more. The resolution in 2016 gave us complete control of the PABC and allowed us to grow faster, further, and without needing to constantly appease and compromise. My hope for the future was that it would allow us to work with the college on well-defined projects such as internships, seminars, and research, but without a need for them to feel responsible or at risk for anything greater than whatever was being collaborated on.

There were multiple reasons for the construction delays. First, because of our conflict with Del Val, which was the most vexing. Then there were self-inflicted problems. Changes we thought would be either welcomed or at worst ignored by the EDA were determined by them to be substantial and require their approval. The first change we proposed was moving the site of the building from one place on our campus to the other. The location we used in the grant application was hastily selected because there were existing township approvals and general designs inherited from the previous building owners. As I mentioned, we grabbed these because, ironically, they expedited our application to the EDA. However, now that we had full control of the project, we thought a better location would be in between our existing buildings, where it would provide a bridge. This made a great deal of sense because the auditorium and cafeteria should be in a central location on the campus.

We prepared new designs, which also included the elimination of the third floor because of height restrictions from the township and the nearby local airport. We are on the flight path and neighbors of the small airport. The EDA took more than a year before eventually agreeing to the change. This brought us to 2018 when the project could start. The actual bids now brought the cost to greater than $17 million, which was without full site preparation. Water runoff concerns dictated that we build a $1 million green roof since we were exceeding our total impervious surface allowance. The township was extremely cooperative and sympathetic, but there were state environmental regulatory requirements. To cure this problem, I suggested that we purchase five acres of a neighboring, undeveloped property, which would then bring our impervious surface total to within the allowable amount. Buying this (for ~$840,000) delayed us another 6–8 months. Project costs continued to rise. In the meantime, we learned that the ground under which the new building would be built was very difficult to work with, being largely rock. There went the lower level, which was to accommodate some of the loss of a third floor, and more approvals were needed.

More delays. More costs. More grants, too. The building was now down to 38,000 ft^2 with two, not three floors. Total project costs reached $21 million. We were able to obtain almost half from the EDA and another $5 million from the State of Pennsylvania. There was broad enthusiasm for us, both federally and locally. Construction finally started with some fanfare in October 2019. The bulldozers and diggers were suddenly all over our campus. There were dozens of construction workers moving piles of materials and a mini-city of temporary trailers. The place looked serious. At long last, this building was going to happen. Who could know what was coming next?

In December 2019, after surviving litigation with Del Val and thinking it was now going to be smooth sailing, the EDA suddenly suspended funding for the project. They said we hadn't kept them in the loop, or some other reasons I still do not understand. They eventually fully reinstated the funding, but not after weeks of answering their questions. One big concern was apparently that the EDA was worried we would not complete the project by the ultimate end date of the EDA funding period, which was October 2021. After that date, they would be unable to release any more funds. We had asked for a number of allowances from them in the past, and these requests, which were reasonable and intended to improve the project, were always eventually granted. However, each cycle of request and approval was eating up time, and there was apparently an absolute statutory requirement that these EDA funds be completely spent and the building be occupied by October 2021 or suffer penalties such as repaying most of the grant funds. Even with their curt comments and delays, completion by October 2021 should not have been a problem.

Then, the COVID-19 pandemic struck. There was concern across the country that public gatherings should be stopped to halt the spread of the virus. Restaurants, shops, and most businesses had to close temporarily, at least until the nation could determine how to control the risks of contagion. In March 2020, the Governor of Pennsylvania issued an order that halted our construction. With the clock ticking toward October 2021, we had another reason to panic, over and above the worries about COVID-19. The EDA was not willing to give us another break on the deadline.

Within a few weeks, the state had permitted us to resume work, but only after special COVID-19 protocols were written and approved by state health authorities. These protocols required that we have a registered nurse onsite every day to screen construction workers, in the morning and end of the day. My wife Joan, who is an RN, stepped in and managed a small team of our master's students (who were also EMT trained). Joan volunteered her time and enabled us to avoid further delays. Even so, the COVID-19 restrictions and new problems getting supplies and construction materials for the project added hundreds of thousands of dollars to our costs.

I like to think that this was the building that insisted on being built, no matter what tried to get in its way. Legal battles. Cost over-runs. A global pandemic. But the building did get built. We obtained a technical occupancy permit in October 2021, thus satisfying the requirements of the EDA grant. The formal grand opening and ribbon-cutting ceremony was held in May 2022 to assure pleasant weather for the ceremonies, but the building was already fully occupied and fully rented by December 2021.

References

Cebolla-Boado H, Nuhoglu Soysal Y (2023) Selectivity among educational migrants? A multisited investigation. Chin Sociol Rev 55:1–25

CEOWORLD (2021) https://ceoworld.biz/2021/01/03/worlds-most-entrepreneurial-countries-2021/

Echeverria-Estrada C, Batalova J (2020) Chinese immigrants in the United States. Migration Policy Institute

Iloeje UH, Yang HI, Jen CL, Su J, Wang LY, You SL, Chen CJ (2007) Risk and predictors of mortality associated with chronic hepatitis B infection. Clin Gastroenterol Hepatol 5(8):921–931

Kostova T, Roth K, Dacin MT (2008) Institutional theory in the study of multinational corporations: a critique and new directions. Acad Manag Rev 33(4):994–1006

Peters BG (2019) Institutional theory in political science: the new institutionalism. Edward Elgar Publishing

Zhang BZ, Haimson OL, Thomas M (2022) The Chinese diaspora and the attempted WeChat ban: platform Precarity, anticipated impacts, and infrastructural migration. Proc ACM Hum-Comput Interact 6(CSCW2):1–29

Chapter 2
Biotechnology and Entrepreneurship Need Business Incubators

Entrepreneurs and insane people are similar. They both see things that don't exist.
Anonymous

This section of the book discusses the life sciences business incubator model, ranging from a discussion of the incubator model to the nature of the startups within them. We then take an in-depth look at the Pennsylvania Biotechnology Center (PABC) as an example of a successful life sciences accelerator/incubator and one about which we have particular knowledge. Business incubators, however, depend on entrepreneurs and their startup companies. In this section, we begin with a look at that.

2.1 Entrepreneurism and Innovation

Entrepreneurs have been called people "who see things that do not exist." Of course, that is also a description of an insane person. Entrepreneurs can certainly be characterized by extreme, bold, risk-taking behaviors that will strike some as irrational. A more sympathetic definition of an entrepreneur is provided in *Webster's Dictionary* as someone who "has an idea and or develops an idea into a product and has a reasonable plan to accomplish the goal." That is nicer.

By definition, innovation is essential to progress. Think about how difficult it would be to drag a heavy object without a wheel. Someone had to think of that, and apparently things were just dragged out before an ancient Mesopotamian figured this out. A more recent and dramatic example of innovation is the COVID-19 vaccine. The first human case of what turned out to be COVID-19 was reported to be in Wuhan, China, in December 2019. Deployment of a safe and effective vaccine occurred in fewer than 18 months. Innovation, on demand, basically saved the world. However, there needed to be a preexisting innovation infrastructure with public and private cooperation for this to all happen so quickly.

Moderna and Pfizer-BioNTech were the producers of COVID-19 mRNA vaccines. Moderna is a relatively new biotechnology company that is developing messenger RNA (mRNA) as a means to express therapeutic proteins and vaccines. Pfizer, working with the startup BioNTech, began to develop COVID-19 vaccines delivering mRNA that produces the viral "spike" protein. The mRNA vaccine approach was made possible by the pioneering work of Drew Weissman, M.D., Ph.D., and Katalin Kariko, Ph.D., both from the University of Pennsylvania. They received the Nobel Prize in Medicine in 2023 for this work.

Briefly, mRNA is naturally encoded in the DNA of all cells and is translated into the protein product it specifies via ribosome complexes in the cell. However, mRNA injected (or transfected) into the cell from the outside is detected by the cell as foreign (not from the cell) and induces a host alarm system that can cause cell death. Drs. Weissman and Kariko found a way to modify the nucleic acids of the mRNA such that they are no longer recognized by the host alarm system and can be used to produce anything the molecular biologist dials in to them. Moderna and Pfizer-BioNTech dialed in mRNA specifying COVID-19 proteins, but it is now clear that one can pretty much program the mRNA to produce anything.

By March 2020, the number of COVID-19 cases began to climb exponentially in the U.S., although the number of deaths was *only* in the thousands. On March 17, 2020, only 3 months after the sequence of the COVID-19 virus genome was reported, Moderna began the first clinical trial of its vaccine. Pfizer-BioNTech's vaccine was also in trials. There were numerous other companies in the race. Some large companies, such as Astra-Zeneca, were working with the University of Oxford on an adenovirus vector approach, and other smaller companies, such as Novavax, were working on a subunit approach. It was unclear if any of the approaches would be effective.

The number of COVID-19 deaths in the U.S. exceeded 300,000 by December 2020, and an anxious nation along with the rest of the world awaited the first results of the Moderna and Pfizer-BioNTech vaccine trials. Before the end of December 2020, less than a year since the identification of the etiological agent of COVID-19 and less than 11 months from the time its nucleic acid sequence was reported, the FDA approved the use of COVID-19 vaccines (Moderna and Pfizer-BioNTech) based on successful clinical trials. Within 10 days, more than a million doses of vaccine were used.

A grateful world was able to literally breathe a sigh of relief and could finally imagine the end of the terrible, at one time feared, civilization-upending pandemic. This was perhaps the most dramatic example, ever, of how innovation, science, and the public and private sectors can be mobilized to solve enormous global problems.

Therefore, yes, innovation in art, science, medicine, and business has been generally regarded as good, if not essential, to civilization and improving quality and length of life. Although there has been some decrease in trust since the COVID-19 pandemic (Kennedy et al. 2022), the Pew Foundation Surveys as well as a national survey posted by the National Science Center for Science and Engineering Statistics indicate that 92% of the U.S. adult population agree that science generates opportunities for the next generation, and 82% believe that the government should fund basic research that leads to new discoveries. These results have remained steady since 2001. A bit of humility is needed, here, however, because that same survey showed that ~40% of the population found that technology was moving too fast, and only 40% had high confidence in the scientific community. The schism between confidence and belief in the value of discovery may explain some of the public discomfort and distrust seen during the COVID-19 pandemic. Sooner or later, there will be a return to acceptance, if not trust, in the scientific community and recognition of the need for innovation.

If society wants innovation, it will need entrepreneurism. Innovation and entrepreneurism go hand in hand, with entrepreneurism as the major vehicle by which innovations are provided. Small new companies (Newcos) are the vehicle by which many entrepreneurs with relatively limited means can develop their innovations.

A spirit of entrepreneurship is necessary for a startup to even exist. The U.S. possesses all of the components necessary for innovation and entrepreneurship to thrive. This is reflected in rankings in which entrepreneurial activity and environment have actually been quantified. There are several ways in which a nation's entrepreneurship has been measured, but almost all metrics used place the U.S. at the top. For example, the Washington, DC-based Global Entrepreneurial Index (GEI) creates regular entrepreneurial scores for each of 137 countries. They use a combination of parameters, including "risk capital available, competitiveness, innovation, skills and stability" (see Table 2.1). In 2019, the U.S. ranked number one, with a score of 83.6. The U.S. has been consistently top ranked throughout the years of the GEI surveys, while other countries have varied in their rank. In 2019, Switzerland, Canada, and Denmark came in just behind the U.S., with scores of 79–82. After Switzerland, the UK was the next top-ranked European nation, with Iceland, the Netherlands, and Ireland following. China with a score of ~41 and ranking about 40 to 41 out of 137 countries, and Russia with a score of 28 and ranking 79, putting it in the neighborhood of Gabon and Georgia, which ranked very low. Uganda and Malawi scored the lowest, at 13 and 14.

CEOWORLD magazine in 2020 surveyed thousands of readers and created an entrepreneurial index that used 18 indicators some of which are similar to those used by the GEI. Again, the U.S. came out on top, although the rankings of other countries differed somewhat from the GEI. In the CEOWORLD ranking, the UK and Germany were second and third, respectively, with Israel in fourth place ahead of the United Arab Emirates, which was fifth. Poland is ranked sixth, Spain is seventh, and eighth, ninth, and tenth positions are held by Sweden, India, and France. Australia took No. 11. Canada was 17th, and Switzerland was 20th. The reasons for the difference between the CEOWORLD and GEI rankings were not clear, but the inclusion of startup numbers and statistics may be an explanation.

The leading role of the U.S. in innovation is consistent with the number of therapeutic medicines discovered in the country. Drugs approved for use by the U.S. FDA are usually used, at least eventually, throughout the world. Using FDA approval as a standard, the U.S. is responsible for 61% of approved drugs between 2011 and 2021 (Bioworld Report: Vitaltransformation 2023; O'Loughlin et al. 2023). Taken together, although precise scoring and rankings may differ from survey to survey for any given country, it does appear that the U.S. is consistently considered a highly innovative and entrepreneurial country, and the fruits of this entrepreneurship are

Table 2.1 2019 Global Entrepreneurial Index[a] (top 20 and bottom 20 countries shown)

Rank	Country	Score[b]
1	United States	83.6
2	Switzerland	80.4
3	Canada	79.2
4	United Kingdom	77.8
5	Australia	75.5
6	Denmark	74.3
7	Iceland	74.2
8	Ireland	73.7
9	Sweden	73.1
10	France	68.5
11	Netherlands	68.1
12	Finland	67.9
13	Hong Kong	67.3
14	Austria	66
15	Germany	65.9
16	Israel	65.4
17	Belgium	63.7
18	Taiwan	59.5
19	Chile	58.5
20	Luxembourg	58.2
107	Honduras	18.7
108	Guatemala	18.5
109	Kenya	18.4
110	Ethiopia	18.3
111	Suriname	18.1
112	Lao PDR	17.8
113	Cambodia	17.6
114	El Salvador	16.7
115	Tanzania	16.4
116	Guyana	16.4
117	Gambia, The	16.1
118	Mali	15.9
119	Liberia	15.7
120	Pakistan	15.6
121	Cameroon	15.4
122	Nicaragua	14.7
123	Angola	14.4
124	Mozambique	14
125	Madagascar	14
126	Venezuela	13.8
127	Myanmar	13.6
128	Benin	13.3

(continued)

Table 2.1 (continued)

Rank	Country	Score[b]
129	Burkina Faso	13.2
130	Guinea	12.9
131	Uganda	12.9
132	Sierra Leone	12.3
133	Malawi	12.2
134	Bangladesh	11.8
135	Burundi	11.8
136	Mauritania	10.9
137	Chad	9

[a]From: The Global Entrepreneurship and Development Institute (GEDI) Washington, DC
[b]The GEI score is based on "entrepreneurial attitudes, risk capital available, competitiveness, innovation, skills and stability …" according to the GEDI website

manifested in drug discovery, among other things. Furthermore, entrepreneurship is the goal of biotechnology business incubators.

2.2 Small Business as a Vehicle for Innovation

Small businesses are both a product and a physical manifestation of entrepreneurship. They are also a critical part of the U.S. economy. The entire U.S. economy, not just the therapeutic drug development pipeline, depends on small companies. Small businesses, defined by the U.S. Small Business Association (SBA) as having fewer than 500 employees, are a major employer in the U.S., with more than 61 million workers in 2018. Almost 20 million of these individuals work for companies with fewer than 20 employees, accounting for 90% of small businesses (U.S. Small Business Administration Small Business Profile 2020). Most small businesses are truly small. Seventy-eight percent of them have fewer than 10 employees (Small Business Trends 2020).

In the technology and biotechnology sectors, small businesses are a major vehicle, if not the source, of innovation. This is reflected in patent applications and drugs reaching approval. Using patent applications as a metric, small businesses "produced 27 patents per 100 employees, compared with 1.6 patents per 100 employees in large firms" (Giglio and Micklus 2021; Kneller 2010; Small Bus Profile 2020). In technology and biotechnology, these patents and discoveries tended "to be more significant than those from large firm patents, outperforming them in a number of categories including growth, citation impact, and originality" (Breitzman 2008; Breitzman and Hicks 2008; Hicks et al. 2000, 2001).

In the area of biotech, small-company therapeutic drug discoveries are more likely to be "non-follow-on" (unique), and many are considered breakthroughs (Kneller 2010). Examples of some of the most notable innovations are the COVID mRNA technologies from Moderna and Pfizer-BioNTech, polymerase chain

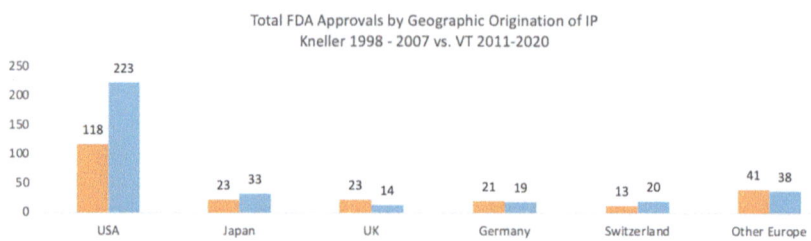

Fig. 2.1 Number of U.S. FDA Drug approvals per year and country of origin. (Source: vital-transformations web site https://bio.news/health/small-biotech-companies-develop-most-new-drugs-in-united-states-vital-transformation/)

reactions (PCR) from Cetus (now a part of Roche), therapeutic monoclonal antibodies from Centocor (now a part of Johnson & Johnson), recombinant non-antibody proteins (i.e., insulin, interferon, and clotting factors) as therapeutics from Biogen, Genentech, and the first reliable oral cure for hepatitis C from Pharmasset (acquired by Gilead Sciences). All are examples of breakthroughs from small biotech companies.

Where are these innovations coming from? The same places are the homes of entrepreneurship. In 2010, Kneller reported that >60% of all the drugs in use (at that time) had originated from the U.S., and > 50% of those had come from universities or small companies. That trend has continued (Kesselheim et al. 2015a, b; Hwang et al. 2018). After the U.S., and along these lines, Japan and the UK were each responsible for approximately 10% of new drugs approved. In contrast, in Japan, almost all of those drugs are from their large pharma sector, with very few from universities. Figure 2.1 graphically illustrates this point.

In the U.S., the contributions of universities and small companies to new drug approvals parallel the growth of recombinant molecular biology, becoming much more significant after 2000 than before (Munos 2009). It is tempting to conclude that there is something of a cause and effect. For example, between 1950 and 2000, the number of new molecular entities (NMEs) and drugs approved by the U.S. FDA (~20–23) per year was fairly constant, with 1 year of exception in 1996 when a backlog of applications was processed (Munos 2009). However, this number had more than doubled by 2015, concomitant and consistent with the biotechnology revolution at universities, the growth of small biopharma, and substantial changes in federal legislation regarding technology transfer, the Orphan Drug Act and mechanisms of approval (i.e., Prescription Drug User Fee Act [PDUFA] in 1992). For example, between 1950 and 1995, "small" pharma was responsible for a steady, ~ 20–25% of NMEs (drugs) approved (Munos 2009). As mentioned, by 2020, this figure doubled to 50–55% (Kneller 2009; Bionews, 12/6/22). Some of the notable medicines and medical devices that were discovered or initially developed by start-ups before being handed over for commercialization to big pharma are shown in Table 2.2. This list, which mentions just a few examples, is impressive.

2.2 Small Business as a Vehicle for Innovation

Table 2.2 Selected top selling/medically important medicines/advances originating in small companies

Advance	Use	Company/Sponsor	Year (discovered / approved for use)
PCR (polymerase chain reaction)	DNA synthesis and detection	Cetus	1983
MRI (Medical Magnetic Resonance Imaging)	imaging	U Aberdeen, UCSF, and Fonar corp	1975–1977
DNA and protein sequencing	research	Genco/Applied Biostystems	1981
Epogen/Epoetin	Dialysis, red cell aplasia	Amgen	1989/1995
Lamivudine	HIV/HBV	McGill/Yale/Emory/Biochem Pharma	1989
Neuopogen/Filgastrin	Neutropenia	Amgen	1991
Humilin/Insulin	Diabetes	Genentech	1997
Avonex/IFb	MS	Biogen	1997
Protropin/Somatotropin	Growth disorder	Genentech	1997
Rituxan/RituximaB	Non Hodgkins	Genentech	1997
Remicade/Infliximab	R.Arthritis/Crohns	Centecor (J/J)	1998
Intron A/IFN	HCV, MS	Biogen	2001
Eylea (aflibercept)	VEGF inhibitor; mac degeneration, cancers	Regeneron	2001
Tyvaso, Remodulin, and Orenitram	Pulmonary hypertension	United Thera	2002
Zavesca	Gaucher/Lysosomal storage	Actelion/Cellgene	2003
Emtricitabine	HIV	Triangle (Emory)	2003
Amyvid®	Radiodiagnostic for amyloid plaques	AVID Radiopharma	2011
Sofosbuvir	HCV	Pharmasette	2013
Trulance	Idiopathic constipation	Synergy Pharma	2017
Spikevax	COVID vac	Moderna	2021
Regen Cov	COVID vaccine	Regeneron	2022
Ciltacabtagene autoleucel (Carvykti) cilta-cel	CART for Multiple Myeloma	Janssen (Legend Biotech)	2022
Covavax/Nuvavoid	COVID vaccine	Novavax	2023

This is not to say that big pharma isn't innovative or essential to therapeutic drug and medical device discovery. Of course they are. First, they continue to be a major source of innovation, with, as mentioned, at least half of all new therapeutics originating with them, even today. However, on a per-employee basis, or when considering breakthroughs, they are clearly less efficient than smaller companies. However, even for most of the smaller companies, large pharma is needed for the costliest parts of drug development. This includes the advanced preclinical and clinical phases (which is not trivial and the place where most investigational therapeutic drugs are disqualified and fail) and, of course, for production and sales. These are the multi-hundred million dollar, even billion dollar, costs that are difficult, if not impossible, for most small companies to shoulder.

The relationship between small and large pharma is, therefore, an essential part of the biopharma ecosystem. It continues to thrive today with partnerships as the usual path by which small biotech can launch a product. These deals are the lifeblood of the field. In 2020, alone, there were at least 68 deals, totaling more than $130 billion in value (Grabow 2023 report). Tables 2.3, 2.4, 2.5 itemize major deals from 2020 to 2022 between small and large pharma.

Table 2.3 2022 Notable deals between big pharma and smaller companies[a]

Big pharma	Smaller pharma licensor	Product/technology	Deal size	Up front (if known)	Date reported
Amgen	Horizon	Many approved drugs	27.8 bil	acquisition	12/22
Merck & Co.	Kelun-Biotech[a]	Preclinical ADC (antibody-drug-conjugates) assets for Oncology	9.4 bil	175 mil up front;	12/22
Kite (Gilead)	Tmmunity	CART for cancer	Undisclosed		12/22
Sumitovant	Myovant	Orgovyx (prostate ca, approved), Myfembree (approved, woman's health)	2.9 bil	1.7 bil	11/22
Pfizer	Global Blood Thera	Oxbryta (voxelotor)sickle cell	5.4 bil		10/22
Novo Nordisk	Forma thera	Developing sickle cell Rx	1.1		9/22
Amgen	Chemocentrix	Tavneos, approved for severe active anti-neutrophil cytoplasmic autoantibody (ANCA)-	3.7		8/22
GSK	Sierra Oncology	JAK inhibitor momelotinib for myelofibrosis	1.9		7/22
BMS	Turning Point Thera	Repotrectinib, ROS1 and TRK inhibitors, cancer	4.1 bil		6/22
GSK	Affinivax	Vac platform	2.1		5/22
Sanofi	IGM Biosciences	Engineered IgM programs for oncology, autoimmunity, inflammation	6.5 bil	150 mil up front	3/22
Biocon Thera	Viatris	Biosimilars	3.0	2 bil	2/22
Pfizer	Biohaven	Nurtec (Rimegepant, FDA approved), other, migraines	11.6 bil		11/21

[a]Note: these data are from public announcements and reports found on the internet

2.2 Small Business as a Vehicle for Innovation

Table 2.4 Smaller strategic deals in 2022[a]

Big pharma	Smaller pharma	Product	Up-front cash ($ millions)	Comments and total (if known)	Date reported
Gilead Sciences (licensor) Dragonfly Therapeutics	Dragonfly Therapeutics (licensee)	collaboration to develop natural killer cell engagers in oncology and inflammation	300	Gilead Sciences (licensor) Dragonfly Therapeutics (licensee)	5/22
Pfizer	Beam Thera	collaboration to advance mRNA in vivo base editing for rare diseases	300	1.5 bil	1/22
Merck & Co.	Orion	commercialization of ODM-208, Ph. 2, an resistant prostate cancer	290		7/22
Cullinan Oncology	Taiho Pharma	joint development of Ph 1//2A TAS6417, lung cancer EGRR inhibitor	275	400	5/22
Emergent BioSolutions	Chimerix	Tembexa (brincidofovir), FDA-approved smallpox drug	238		9/22
Merck & Co.	Orna Therapeutics (licensor)	vaccines and therapeutics using Orna's oRNA-LNP technology	150 (up to $3.5 bil)	Orna Therapeutics (licensor) Merck & Co. (licensee)	8/22
Roche	Repare	CRISPR CAS, camonsertib (RP-3500), Ph. 1/2	125	Roche (licensee) Repare Therapeutics (licensor)	6/22
Roche	Poseida Therapeutics	Poseida grants Roche exclusive rights to develop and commercialize allogeneic CAR-T cell therapies for hematologic malignancies	110	Roche (licensee) Poseida Therapeutics (licensor)	8/22
Bristol Myers Squibb	Century Therapeutics	collaboration to develop induced pluripotent stem cell-derived allogeneic cell therapies.	100	Bristol Myers Squibb (licensee) Century Therapeutics (licensor)	1/22
Sanofi	Exscientia	collaboration to develop artificial intelligence-driven pipeline of precision-engineered medicines	100 (up to 5 bil)	Exscientia (research partner) Sanofi (research partner)	1/22

(continued)

Table 2.4 (continued)

Big pharma	Smaller pharma	Product	Up-front cash ($ millions)	Comments and total (if known)	Date reported
Shionogi	F2G, Ltd.	Shionogi and F2G enter strategic collaboration to develop and commercialize antifungal agent olorofim in Europe and Asia	100	F2G (licensor) Shionogi (licensee)	5/22
Merck	PMV	PC14586 is a first-in-class, small-molecule p53 reactivator	Clinical collab		6/22

[a]Note: These data are from public announcements and reports found on the internet and Micklus and Giglio (2020), Giglio and Micklus (2021), Redit (2022)

Table 2.5 Notable deals between big pharma and smaller companies (2020/21)[a]

Big pharma	Smaller pharma licensor	Product/technology	Deal size	Up front	Date reported
CSL	Vifor	Pipeline for iron deficiency (Ph.2, VIT-2763) and cardio	$11.7 Bil	All cash	12/21
Merck	Acceleron	Sotatarcept Ph. 3 Pul Hypertension Rx, Reblozyl (Approved, anemia)	$11.5 Bil		9/21
Jazz	GW Pharma	Epidiolex, seizures approved	$7.2 Bil		12/21
Amgen	TeneoBio	Heavy Chain Ab platform; TNB 585, Ph. 1 Prostate Ca	$2.5 Bil	$900 M	7/21
Horizon	Viela	Upliza, approved for neuromyelitis optical spectrum	$3 Bil		3/21
Sanofi	Kymab	Ky1005, mAb to OX40-Ligand (immune)	$1.45 Bil	$1.1 Bil	4/21
Amgen	Five Prime	Emarituzumab, anti-FGFR2b antibody Phase 3 Gastric Cancer	$1.9 Bil		3/21
Gilead	Myr	Hepcludex, HDV, Ph. 3	$1.5 Bil	$1.3 Bil cash	10/20
Merck	Seattle Genetics	Liv1 mAb for breast cancer	$4.5 Bil	$1.6 Bil	9/20
Biogen	Denali	LRRK2 inhibitors for parkinsons	>$2 Bil	$560 M	8/20
Merck	Astex/Taiho	Kras inhibitors, cancer	$2.5 Bil	$50 M	1/20
Abbvie	Genmab	Bi-specific Abs for cancer	$3.9 Bil	$750 M	6/20
Biogen	Sage	Zuranolone for multiple depression types and SAGE-324 for essential tremor	$3.1 Bil	1.5 Bil	11/20
Biogen	Sangamo	Neuro program	$2.72 Bil	$350 M	2/20
Sanofi	Nurix	Protein degradation	$2.55 Bil	$55 M	1/20

[a]Note: these data are from public announcements and reports found on the internet; *M* Millions, *Bil* Billions

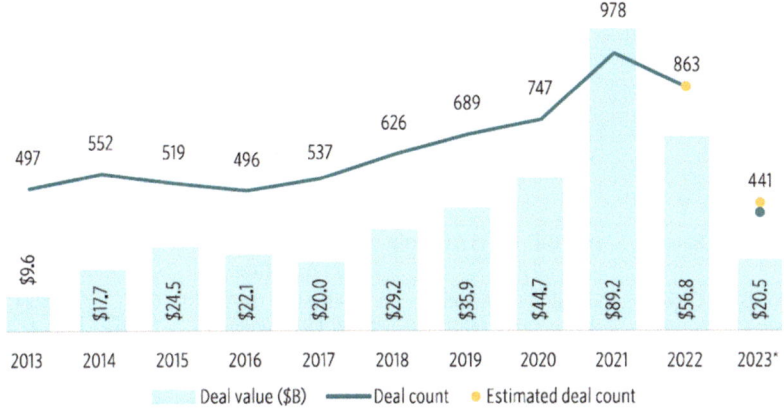

Fig. 2.2 Deals as venture investments (U.S.) 2013–2023. (Source: Pitchbook 2023)

It is interesting to note that many deals occurred for small biotechs that had only preclinical or very early clinical stage assets. Additionally, there is something of a trend toward biologicals.

Global economic trade and geopolitical upsets in 2022, such as Russia's invasion of Ukraine, are generally blamed for fewer deals than in 2020 and 2021. The 2022 total was less than 80 and an amount less than $60 billion, and there were 21 IPOs in 2022, compared to 111 in 2021 (Pitchbook 2023). See Fig. 2.2. That said, there continued to be several major deals providing bright spots for an otherwise uncertain year in the national economy and things seemed to be picking up at the end of the year. (See Tables 2.6, 2.7) and pent up un-invested capital will put pressure on fund managers to make investments in the coming year (see Table 2.8).

2.3 Startup Company Anatomy

The Edison Electric Light Company may be the first U.S. startup company founded with venture capital. Established in 1878 in Schenectady, New York, by Thomas Edison and Charles Coffin, its funding came from J. P. Morgan. It received additional funding from the Drexel Morgan and Company and merged with Thompson Houston Electric Company of New Britain, CT, and became the more familiar General Electric Company in 1892. Therefore, startups, in the way we think of

Table 2.6 Biotech deals in the first quarter of 2024[a]

Acquired	Acquired by	Value ($)[b]	Indication[c]	Lead asset[d, e]	Development status[f]	Q1 2024[g]
IFM Due	Novartis	$90 M/ $835 M	Chronic dis/ innate immune stim	STING a[e]	PC	3/24
CymaBay	Gilead	$4.3 Bil	PBC (liver)	Selapar (PPARd a[e])	NDA	2/24
Morphosys	Novartis	$2.9 Bil	Fibrosis/Onco several	Pelabresib (BETi[e]) & Tulmimetostat (EZHi)	III	2/24
Inhibrx	Sanofi	$1.7 Bil	A1ATrypsin	INBRX101 (FcAAT fusion)	I	1/24
Aiolos Bio	GSK	$1 Bil	Asthma	AIO-001 (mAb[e] to TSLP)/others	I	1/24
Harpoon Thera	Merck	$80 M	Oncology (lung)	MK6070 (T cell engager prot)	I/II	3/24
Calypso	Novartis	$250 M/ $470 M	GI/ Celiac	CALY002 (mAb[e] to IL15)	IB	1/24
Mirata	BMS	$4.8 Bil	Multiple Oncology	MRTX1719 (PRMT5 i[e]), others	I & II	1/24

[a] Major Deals involving US companies shown. Information from Company press releases and https://www.biopharmadive.com/news/biotech-pharma-deals-merger-acquisitions-tracker/604262/
[b] Deal dollar amount with milestone upside shown where known. *M* Million, *Bil* Billion
[c] Disease area targeted by the asset. Major area reported where multiple were reported
[d] Lead asset. Name of asset and general mechanism
[e] Abbreviations used in asset column: *a* agonist, *I* inhibitor, *R* receptor, *mAb* monoclonal antibody
[f] Phase in development. *PC* preclinical; Clinical Phases I, II, III; *NDA* New Drug Application filed; Approved indicates US FDA approval (or European if indicated); In Use means approved and marketed
[g] Date of press release announcement

them, with venture funding and mergers, are a relatively recent part of civilization and arrived concomitantly with the industrial age of innovation.

Startups are perhaps the clearest example and expression of entrepreneurship in a capitalistic country. The Small Business Administration (SBA) defines a startup as a "new business that hires an employee for the first time." A more descriptive definition that I prefer considers entrepreneurship and is used by StartupRanking.com. It defines a startup as a company less than 10 years old with "high innovation." With that definition, they published a list of countries with the most startups. The U.S. with >76,000 has by far the most startups in the world, and this is consistent with its top status in entrepreneurship. The UK, with ~6,200, is a distant second, and Canada, with 3,200, is a distant third (firstsiteguide.com/startup-stats and startupranking.com/countries). All other countries fall well below those. China, as an example, has fewer than 700. Ranked by both total number of startups and startups per million

2.3 Startup Company Anatomy

Table 2.7 Biotech deals of 2023[a] (last quarter 2023 shaded)

Acquired	Acquired by	Value ($)[b]	Indication[c]	Lead asset[d, e]	Development status[f]	Q1 2024[g]
Karuna Thera	BMS	$14 Bil	Neuropysc	MuscarRa	NDA	12/23
Gracell	AstraZeneca	$1 Bil	Cancer	Several CART	I,IIB	12/23
Icosavax	AstraZeneca	$800 M	RSV vaccines	Meta pneumonia vaccines	III	12/23
Cerevel Thera	Abbvie	$8.7 Bil	Neuropysc	Several	I, II,III	12/23
Carmot Thera	Roche	$2.7 Bil	Diabetes/Obesity G	GLPa	II	12/23
ImmunoGen	AbbVie	$10 Bil	Ovarian, other cancers	ADCs	I, III	12/23
RayzeBio	BMS	$4.1 Bil	Cancers	Radiopharma	I, III	12/23
Propella Thera	Astellas Pharma	$175 M	Prosta Ca/androgen bio inhib	Andro bio inhib	I	11/23
Forge Biologics	Ajinomoto	$554 M	Gene Therapy	Platform	I/II	11/23
Telavant	Roche	$7 Bil	Inflam Bow Dis, Fibrotic	mAbs	II	10/23
Mirata	BMS	$4.8 Bil	Onocology	PD1 inhib, others	I, Approved	10/23
Mitokinin	AbbVie	$110 M/$545 M	Parkinsons	PINK1 activator	PC	10/23
Orchard Thera	Kyowa Kirin	$387 M	Metachromleukodystr	Gene Rx	Approved/Europe	10/23
Point Biopharma	Eli Lilly	$1.4 Bil	Oncology	Radiopharma	III	10/23
Alfasigma	Intercept Pharma	$749 M	PBC	FXr-a, Ocoliva	Approved	9/23
Mindset Pharma	Otsuka Pharma	$59 M	CNS	S-5HTR-a	PC	8/23
Zynerba Pharma	Harmony	$60 M	CNS, Fragile X	Cannabinoid patch	Zygel, Approved	8/23
Decibel Thera	Regeneron	$107 M	Hearing loss	AAV gene Rx	I/II	8/23
Reata Pharma	Biogen	$7.3 Bil	CNS, Rare Dis, Friedreichs Ataxia	Skyclary	Approved	7/23

(continued)

Table 2.7 (continued)

DTx Pharma	Novartis	$500 M	CNS, Rare, Charcot Marie Tooth, others	siRNA platform	PC	6/23
Chinook	Novartis	$3.2 Bil	Nephropathy: Oral endothelial A & mAb	Atrasentan & Zigaibart	III (2 assets)	8/23
Prometheus	Merck	$10.8 Bil	GI, Crohns, UC/TNFa	mAb (MK7240)	II	6/23
Paratek Pharma	Gurnet Point Cap	$123/462 M	Antibiotics for pneumonia, other	Nuzyra (Omadacycline)	In use	6/23
VectivBio	Ironwood Pharma	$1.14 Bil	Rare GI Dis (short bowel, aGVHD)	GLP2a/aproglutide	III	5/23
CTI Biopharma	Sobi	$1.7 Bil	Oncology/IRAK1 kinase inhib	Vonjo (pacritnib)	Approved	5/23
Iveric Bio	Astellas Pharma	$5.9 Bil	Vision/Optham	Izervay (intravit sol) & others	Approved & I,II assets	4/23
Spectrum Pharma	Assertio Holdings	$248 M	Febrile neutropenia/injection	Rolvedon	Approved	4/23
Bellus	GSK	$2 Bil	Chronic cough	Camplipixant (P2X3Ra)	III	4/23
Seagen	Pfizer	$43 Bil	Oncology/ADCs	Adcetris, Padcev, Tivdak, Tukysa	4 approved	3/23
Concert Pharma	Sun Pharma	$579 M	Alopecia areata	Deurxolitnib oral JAK inhib	III	1/23
InstaDeep	BioNTech	$462 M	AI-driven drug Dis	RiboCytokine, RiboMab platforms	Support dev	1/23
CinCor Pharma	AstraZeneca	$1.3 Bil	Hypertension/Kidney Dis	Baxdrostat (Aldosterone inhib)	II	1/23
Albireo	Ipsen	$952 M	Rare liver diseases/Famil intrahep chlol, 2 rare ped liver Dis	Bylvay (bile acid transp inhib)	Approval for Pruritis, Phase III for Biliary Atresia	3/23
Amyrt Pharma	Chiesi	$1.25 Bil	Rare dis	Several marketed	In use	1/23
Ori Ciro Gen	Moderna	$85 M	Platforms for cell free DNA synthesis	Research tool	N/A	1/23

Date of press release announcement

[a] Major Deals involving US companies shown. Information from Company press releases and https://www.biopharmadive.com/news/biotech-pharma-deals-merger-acquisitions-tracker/604262/

[b] Deal dollar amount with milestone upside shown where known. *M* Million, *Bil* Billion

[c] Disease area targeted by the asset. Major area reported where multiple were reported

[d] Lead asset. Name of asset and general mechanism

[e] Abbreviations used in asset column: *a* agonist, *I* inhibitor, *R* receptor, mAb monoclonal antibody

[f] Phase in development. *PC* preclinical, Clinical Phases I, II, III, *NDA* New Drug Application filed; Approved indicates US FDA approval (or European if indicated); In Use means approved and marketed

2.3 Startup Company Anatomy

Table 2.8 Private equity cash reserves (2014–2023)[a]

Year	Global "dry powder" (trillion US$)	Global M&A deals (trillion US$)	Merger investment/ "dry Powder" (available)
2023	2.49	0.78	0.31
2022	2.24	1.40	0.62
2021	2.26	2.13	0.94
2020	2.22	1.16	0.52
2019	1.84	1.11	0.60
2018	1.68	1.10	0.65
2017	1.52	0.97	0.63
2016	1.25	0.88	0.70
2015	0.971	1.11	1.14
2014	0.902	1.01	1.02

Source: Visual Capitalist (https://www.visualcapitalist.com/sp/2-5-trillion-private-equity-cash-reserves/) which used: Dealogic, S&P Global, Preqin, and Citizens JMP Securities. PEM&A deal value reflects sponsor-related M&A announcements
[a]Global Private Equity Cash Reserve and Merger and Acquisition Deal values in US dollars (estimates)

Table 2.9 The 15 countries with the most start-ups

Country	Number of start-ups[a]	Population[b]	Start-ups per mill people
US	75,688	340,000,000	222
India	15,966	1400,000,000	11.4
UK	6,986	67,000,000	104.3
Canada	3,780	38,200,000	99.5
Australia	2,733	25,600,000	109.3
Indonesia	2,521	273,800,000	9.2
Germany	2,419	83,200,000	29.1
France	1,627	67,750,000	24.0
Spain	1,473	47,420,000	31.3
Brazil	1,182	214,300,000	5.52
Netherlands	1,101	17,530,000	64.8
UAE	1,065	9,441,129	113.3
Israel	986	9.038,000	109.2
Italy	956	59,111,000	16.17
Switzerland	808	8,740,000	92.44

[a]Number of starts-ups from https://www.startupranking.com
[b]Population based on latest census data

people, the U.S. is number one (see Table 2.9). After the U.S., with ~222 startups per million, the UAE, Australia, Israel, UK, Canada, and Switzerland rank second, third, fourth, fifth, sixth, and seventh, respectively, clustering in each with ~100 startups per million (Table 2.9). Of the 76,000 startups in the U.S., ~6.8% and ~ 7% are in health/biotech/life sciences and fintech, respectively. Information technology is also well represented, although most startups are outside of high technology. (firstsiteguide.com/startup-stats). Table 2.10 shows a summary.

Table 2.10 Categories of startup companies in the U.S.

Fintech	7.1
Life science	6.8
AI	5.0
Gaming	4.7
Adtech	3.3
Edtech	2.8
Cleantech	2.1
Blockchain	1.5
Robotics	1.3
Cybersecurity	0.7
Agtech	0.6
Other	64.1

Source: Statistica 2023 https://firstsiteguide.com/startup-stats

The demographics of those who create small-medium enterprises (SMEs) and startups are also revealing. Recognizing that the definition of a start-up company can vary. A study by the Kauffman Foundation used the criteria "venture backed" and a Crunchbase database to analyze 90,000 venture-funded firms. They found that 56% of the executives had at least a bachelor's degree (West and Sundaramurthy 2019). Other studies have claimed that as many as 95% of U.S. startups were founded by people with college education (firstsiteguide.com/startup-stat; the Kauffman Foundation 2014). The average age of a startup founder varies slightly depending on the sector and type of company. Founders of companies based on patents are somewhat older (42–45 years) than founders of digital/computer-oriented companies, where the age ranges from 38–40 years old (Brieger et al. 2021; Azoulay et al. 2020). Founders of manufacturing-based companies were the oldest, at 47–51 years (Azoulay et al. 2020). The age and degree of education of these founders are somewhat contrary to some fashionable thinking that, to be a successful entrepreneur, advanced education is not needed.

Of course, there are some extraordinary and well-known exceptions to this rule, such as the founders of Microsoft, Facebook, and Apple. In addition, the trends may be changing, but not so much in life sciences. In another report from the Kauffman Foundation, "Trends in Entrepreneurship" (2021), and a report from Fairlie and Desai (2020), it was found that the profile of the entrepreneur in 2020 is somewhat different than it was in 1996—2000, but not so much regarding educational level. It was in age. As shown in Table 2.11, the number of older entrepreneurs doubled in representation (although youth is still predominant), and there are many more foreign-born and minority-represented entrepreneurs. Hispanic ethnicity showed the greatest increase. The number of female entrepreneurs remained relatively steady but with a slight decline from ~44 to 39%. Table 2.11 summarizes the findings, which are based partly on statistics from the SBA at the U.S. Bureau of Labor Statistics.

The qualities outlined in Table 2.11 objectively describe the demographics of those starting a new business but do not indicate which, if any, properties are

Table 2.11 How have the demographics of entrepreneurs who start new companies changed over time?[a]

	1996	2006	2016	2021
Gender				
Men	56.3	57.7	60.5	60.1
woman	43.7	42.3	39.5	39.9
Ethnicity				
Asian	3.4	5.1	7.6	7.3
Black	8.4	9.6	9.2	10.1
Hispanic/Lat	10.0	15.0	24.1	24.2
White	77.1	67.2	55.6	54.5
Age				
20–34	34.3	27.1	24.4	26.2
35–44	27.4	24.2	24.0	26.1
45–54	23.5	28.1	26.1	24.9
54–64	14.8	20.6	25.5	22.8
Birthplace				
Foreign born	13.3	20.1	29.5	28.8
US Born	86.7	79.9	70.5	71.8

Source: Fairlie and Desai (2020) and Trends in Entrepreneurship Kauffman Foundation 2021; "New" entrepreneurs (Kauffman Foundation and Bureau Labor Statistics/SBA web site)

associated with success or failure. This is considered from the perspective of companies growing at life sciences incubators in the sections that follow.

2.4 The Business Incubator Concept

This section of the book discusses the life sciences business incubator model, shifting between a discussion of the incubator model and the nature of the startups within it. We then take an in-depth look at the Pennsylvania Biotechnology Center (PABC) as an example of a successful life sciences accelerator/incubator and one about which we have particular knowledge.

Business incubators, which may be considered to be a subcategory of technology parks, are loosely defined as for-profit or nonprofit entities "that help nascent businesses get started" (Nicholls-Nixon et al. 2021; Hassan 2020; INBIA website). They seem to be popping up everywhere, and although it is difficult to find precise numbers, in 2020, the International Business Incubator Association (InBIA) estimated there are more than 1,400 in the U.S. alone, of which ~120 are considered specializing in life sciences (Wesser-Brawner, A. Impact Survey, InBIA, 2018). We have seen estimates as high as 1,600, depending on the definition (InBIA, Haynes et al. 2012a, b).

Are they necessary? Do they matter? Considering the "survival of the fittest" competitive world of the free market, do they do harm by keeping businesses alive that should be allowed to fail? At the other extreme, should there be hospitals to help businesses for the common good that are sick and need a hand? These questions will be addressed in this section.

Although startup companies in the U.S. go back into the nineteenth century, using Thomas Edison's General Electric as a storied example, planned business incubation in "incubators," in which multiple different small, new companies are clustered together and use shared services and resources, is a relatively new phenomenon. In the U.S., it is generally accepted that business incubators began in the second half of the last century in Batavia, NY. In 1956, Joseph L. Mancuso and family purchased a complex of buildings from the Massey-Ferguson Company, an agricultural machinery manufacturer, which was closing. Mr. Mancuso was apparently unable to find a single tenant for the entire very large, >800,000 ft^2 space and, thus, for practical reasons, in 1959 began renting out smaller subsections to multiple tenants. His first tenant needed just 2,000 ft^2. Others were soon added. He also provided various services to the tenants over and above the usual (e.g., electricity, heat, etc.), and thus, by definition of business incubators used today, earned the claim to be considered the first business incubator in the U.S. Remarkably, the incubator thrives today as the Batavia Industrial Center and is considered an important contributor to that region of Western New York's economic development.

The University City Science Center in Philadelphia, PA, claims to be the first life sciences incubator in the U.S. and perhaps the world. With its opening in 1963, this claim appears to be valid. It is a nonprofit organization but now collaborates with several for-profit entities, such as the Cambridge Innovation Center and Wexford Science & Technology, reasoning that this assures its life sciences companies have the facilities and resources they need to grow.

It still functions as a consortium of 31 member universities and research institutions in Philadelphia, which have representation on its board. The members include many of the major research institutions in the city, such as the University of Pennsylvania and Drexel University, both of which neighbor the Science Center, and nearby Thomas Jefferson University and Temple University. In addition, a number of local colleges are members, including Bryn Mawr College, Haverford College, Swarthmore College, Lehigh University, and health care institutions, such as Mercy Health and the Presbyterian Foundation for Philadelphia. It certainly, therefore, brings the near universe of Philadelphia area academic and nonprofit research institutions together as stakeholders. It is also one of the larger economic development organizations focused on startups in the U.S. with a budget of $21,474,000 per year (2020) and staff of approximately 75 people.

To some extent, this was enabled by an important operational change between 2015 and 2017 at the center. Prior to 2015, there were ~174 people working in the 37 "resident" companies in its original multistory lab, office, and instructional space in West Philadelphia (Battelle Memorial Institute 2007). In 2015, the Science Center entered into a joint partnership with developer Wexford Science & Technology (Wexford) to develop the former University City High School site. That site, combined with the Science Center's "legacy campus" on Market Street, was rebranded as "uCity Square." Since then, multiple properties have been developed and folded into the uCity Square identity. It now includes more than 200 science, technology, and healthcare companies ranging from one-person startups to growing

2.4 The Business Incubator Concept

companies collectively with more than 15,000 employees working in 20 buildings that collectively exceed >4 million sq. ft. To illustrate the uCity Square community diversity properties, a sampling of the tenants is listed here:

- 3675 Market Street (office, lab, academic, and retail)
- Drexel Health Sciences Building (academic)
- One uCity Square (office and lab space)
- Anova uCity Square (residential)
- Science Leadership Academy Middle School (K-8 public school)
- 3838 Market Street is the newest development in 2023 (office, lab, and retail)

As explained by Kristen Fitch, Senior Director, Marketing, uCity Square (2022), "Our agreements with Wexford (along with Wexford's capital partner, Ventas) and subsequently Cambridge Innovation Center to manage the wet labs, enables the Science Center to put more resources toward impact programming without compromising the significance of real estate in our operations." The Science Center's mission-driven programming works in conjunction with real estate to support early-stage and growing life sciences companies. The real estate and programming are integrated in a way that supports the life sciences industry.

The Science Center can boast a history of making significant contributions to the region's (and nation's) economy. This of course is helpful in making the case for public support for this incubator as well as others. Of the 350 graduate organizations referenced in a study by Battelle (2007) the following was reported, "… the 93 [companies that graduated and] remain in the region employ 15,512 people; the Science Center's ~ 39 current incubator residents employ another 174. Using 2007 numbers, these highly skilled jobs command an average wage of $89,000, contributing >$22.0 million to the City of Philadelphia in wage taxes and $42.5 million to the Commonwealth of Pennsylvania in income taxes annually."

If the University City Science Center and Batavia Industrial Center were the first incubators, what gets us to the more than 1,400 in the U.S. today? It was not until 1986 that Robert Smilor and Michael Gill published a book, titled "The new business incubator," which could be considered the "standard" of the industry. By this time, the biotechnology revolution was well underway, with recombinant molecular biology having already made quite an impact on applied commercial biology. Smilor and Gill had a finger on the pulse, and their book about business incubators became something of a handbook for life sciences incubators. It was still the early days though.

In the U.S., in 1980, there were fewer than 50 incubators. By 1985, there were still fewer than 100 incubators opened or planning to open (Wiggins and Gibson 2003; Tsaplin & Pozdeeva 2017). Some states got off to an early lead in the 1980s. Pennsylvania had the largest reported number, followed by California and Michigan. Minnesota, Ohio, and Illinois each had a few, but interestingly, Massachusetts, which is now an incubator leader, was not dominant. Most of the early incubators were established in adaptive re-used facilities (e.g., old factories, warehouses, and schools). For example, 54% of the incubators were in buildings that were greater than 50 years old, and the average incubator was 40,000 ft^2. The growth in incubator numbers is illustrated in Fig. 2.3a, b.

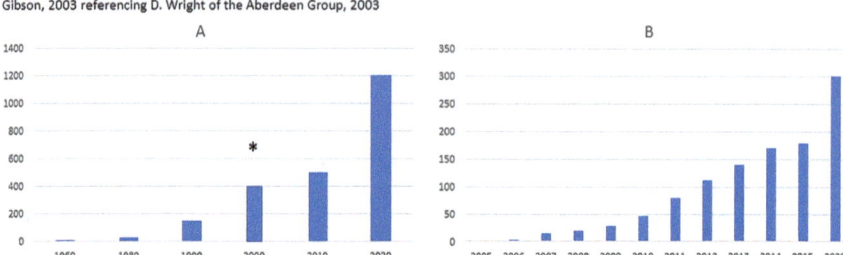

Fig 2.3 Number of Business Incubators and Accelerators. Total number of incubators (**A**) and the subset of business incubator self designated as "accelerators" (**B**) is shown. Numbers of business incubators in the U.S. Source: Startup-insights/startup-growth/how-do-startup-accelerators-work. Brookings Institute 2016 and data for Year 2000 is from Wiggins and Gibson, 2003 referencing D. Wright of the Aberdeen Group, 2003. Number of Business accelerators in the US. Source: Silicon Valley Bank (2020), Brookings Institute (2016), (*) data for 2000 from Wiggins and Gibson reporting results from the D Wright of the Aberdeen Group 2003

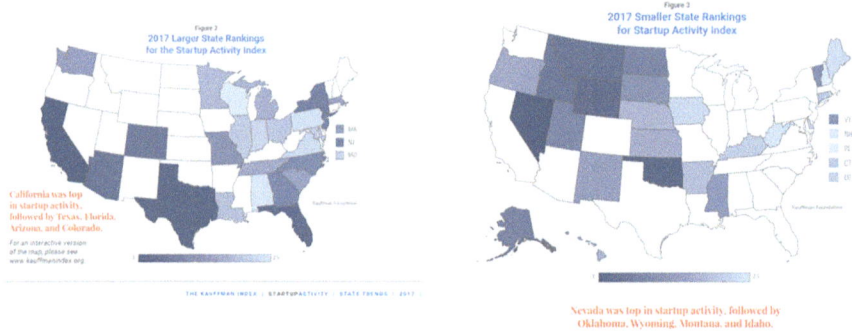

Fig. 2.4 Startup activity by State (**a**) relative amount of start-up activity in larger states (**a**) and smaller states (**b**). (Source and graphic: Kaufman Foundation, Tareque et al. 2017)

Things changed. Paralleling the biotechnology revolution, it appears there was an explosion of incubator growth through the 1990s. Through the 2000s, many incubators were being built in new facilities, and the states with the most incubators had changed dramatically. It has certainly become fashionable for universities and public economic development authorities to create business incubators in or near university campuses and distressed economic zones. By 2020, business incubators had become as common as shopping plazas, with almost every county creating one.

Although getting a precise number is difficult, the InBIA estimated >1,200 in 2020, and this number has apparently shot up to more than 1,400 in 2022. The growth in number over time is shown in Fig. 2.3 with start up geographic distribution illustrated in Fig. 2.4. Two states now have the most incubators: New York has more than 60 (with 17 in bio) and Oklahoma declined to 36 from a high of 48 in 2008. These are followed by California, Ohio, Pennsylvania, Illinois, and Massachusetts. Virginia, Louisiana, Maryland, Minnesota, Wisconsin have between

19 and 35. Idaho, Missouri, Florida, Texas, Kentucky, West Virginia, Georgia, and Alabama have 10–19. All other states had at least one (Tareque et al. 2017; Fairlie & Desai 2020). Farlie and Desai in 2020 determined startup activity indexes for each state (normalizing for state population), and it is interesting to see that this incubator distribution and startup activity generally overlap (See Figs. 2.3 and 2.4).

2.5 Categories of Business Incubators

It is generally accepted that business incubators should provide resources for the entrepreneur that would be difficult for them to obtain on their own. That is, a business incubator should make innovations and business enterprises possible that would be otherwise difficult or improbable. In the U.S., business incubators can be subcategorized as summarized below.

Research Park: These are real estate enterprises with a thematic cluster: an area where businesses involved with science and technology operate. No interrelationships between the businesses operating or the owner and manager of the premises are implied.

Business Incubators: These are facilities that provide space to early-stage companies and assist the growth of new startups by providing a range of services. They nurture new companies, although the ways in which this help is provided varies with the incubator. It ranges from merely providing short-term, flexible leased space to a suite of services, from common infrastructure, equipment, professional services, and even funding. Some are indifferent to company platforms and products, while others are thematically restricted. Some are restrictive in "admission," others less so. Some are nonprofit or university-affiliated, some are publicly owned and run, and others are for-profit enterprises themselves.

Life Sciences Incubator: This is the subcategory of business incubators that are thematically focused on life sciences, human and agricultural medical advances, therapeutics, diagnostics, nonmedical agricultural, and vaccines.

Business Incubator Accelerator: Recently, a subcategory of business incubators has been referred to as business accelerators since they provide exceptional tools, services, and an environment for the entrepreneur tenants over and above providing only space, which is closer to what is provided in a research science park (Bone et al. 2017; Hassan, 2020). The distinction between accelerators and incubators is not sharp, but in general, using the definitions from the literature, accelerators provide nurture to the newest, closest-to-concept startups. Some studies suggest that accelerators were introduced in 2005 with the opening of Y Combinator in Cambridge, MA, and Mountainview in CA, and the PABC, which began in 2006. The number of accelerators shot up after 2010 (Fig. 2.3b, 2.4b). But Y Combinator set the standard. In addition to space, as is provided in the incubator, other resources and services critical to nurturing new startup companies are provided. Acceptance into the accelerator is selective; the services pro-

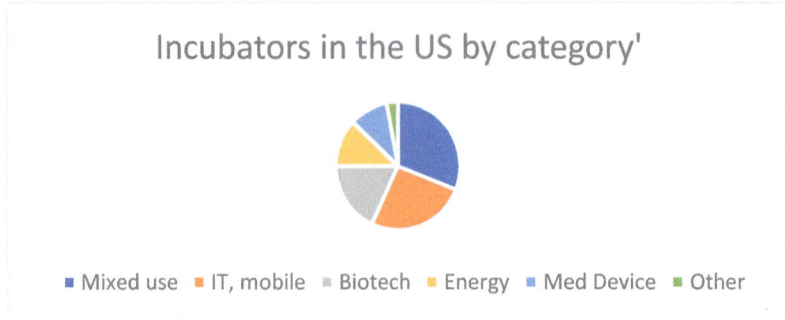

Fig. 2.5 Categories of business incubators in the U.S. (Source: NBIA website)

vided are often considered more important than space; and equity is often taken in lieu of rent. Occupancy in the accelerator is usually for a shorter period of time than in an incubator.

Incubators can be mixed use or thematically focused. Even those that are mixed use can segregate into themes and specialty areas, and it can be seen that only a fraction of the incubators in the U.S. are devoted to the life sciences. A 2016 survey conducted by the InBIA showed the following: The 2016 survey of 248 respondent organizations is shown in Fig. 2.5:

31% (78) Mixed use
26% (65) Internet, mobile, digital
18% (44) Biotech
12% (29) Energy
10% (27) Medical devices.

2.6 Advantages to the Entrepreneur and Community of a Life Sciences Business Incubator/Accelerator

Given the growth in business incubator numbers, it may go without saying that they must offer valuable services to the Newcos that are located within them. Below, we consider the leading advantages they provide to the new companies.

Development Space and Resources: Biotechnology and life sciences startups often need equipment and infrastructures that are extremely expensive and would be beyond reach with the discounted and flexible terms provided. Often, an entrepreneur can begin with access to a single lab bench with access to shared equipment and services, if necessary, making the cost entry threshold within reach of many enterprises.

Shared Equipment and Services: Lab space for the startup's primary and everyday use, lab benches, laminar flow, chemical fume hoods, lab grade sinks, and minor equipment is often provided as a standard and may be within the startup's exclusive space and used regularly. But biotechnology startups often have only an occasional need for very expensive preparative and analytical equipment. Examples in 2023 include ultracentrifuges, high-end preparative HPLC, NMR, mass spectrophotometers, DNA and RNA sequencing, fluorescent activated cell sorters, low-temperature freezers, fluorescent deconvolution, and confocal microscopy. These multi-hundred-thousand-dollar items, which may even require professional staff to operate, can be made available to the single-person startup company in many business incubators.

Access to Science-Oriented Networks: Business incubators often host events that promote networking with other scientific and business professionals.

Access to Nonscience Professional Services: The incubators may maintain lists of nonscience individuals and services and may even offer a menu of professional services to the entrepreneur such as legal, intellectual property, management, accounting, payroll, insurance, joint purchasing, and information technology (IT).

Access to Capital: Many incubators take an active role in introducing the investor community to the startups at their incubators. Startups in many incubators often have privileged access to public money (from state and other municipal programs). Moreover, a growing number of incubators are linked with venture funds, and companies in those incubators have privileged access to these funds.

Benefits of a Knowledge Community: There are cultural benefits to being at an incubator. Many incubators host regular scientific colloquia and business development workshops as well as training programs for the entrepreneur and for students, all of which provide an intellectually stimulating environment. This can be an enormous advantage to entrepreneurs over working in merely rented space in which the real estate owners are relatively uninvolved or indifferent to the tenant. Startups within the incubator can get to know each other, and often joint ventures or complementary activities are initiated that can sometimes be critical to the survival of the new companies.

Legitimacy: In comparison to operating out of one's home or garage, or even a storefront address, having the address of an established institution such as a modern business incubator can help enhance the startup's credibility and reputation. It can improve the ability to attract employees, sponsors (for grant submissions), partnerships, and investors. It can reasonably increase the confidence of these funders and even employees in the legitimacy and soundness of the Newco.

2.7 Defining Success for Life Sciences Incubator

Why do some incubators thrive? Which incubators can be considered successful? Success for an incubator can be simply defined as a generation of new companies that endure. The likelihood that an incubator company would survive for more than 5 years was largely attributed to the selection criteria for admission to the incubator

(Hackett and Dilts 2004). The selection of appropriate tenant companies is mentioned repeatedly as key to the success of the incubator (and the Newcos within it), along with the assistance provided and its availability (Hackett and Dilts 2004; 2008; Fernández et al. 2018) and the suite of services the incubator provides (Fernández et al. 2018). Companies launched from successful incubators had a survival time much longer than those started outside of incubators. For example, more than 50% of new companies started at incubators were still in business after 5 years, whereas only 20–25% of those started outside incubators were around (U.S. Bureau of Labor Statistics; Small Business Association; Shane 2008; Eckhardt & Shane 2011).

Of course, if the primary determinant of the success of an incubator is based mostly on how well it selects its tenants, the value the incubator provides might be narrow. A teacher with excellent, intelligent students can only take so much credit for the student's success. I recall as a beginning faculty member at TJU with a group of others congratulating ourselves on the success of our medical and doctoral student performances on a standardized exam, taking a bow for our teaching. We were pretty good, and TJU was known for its teaching quality. Then someone reminded us that these kids were the cream of the crop and would have done well no matter how good or poor our teaching was. Therefore, if the selection of companies for an incubator is the most important component of their success, then the selection process may be the most important step. I think its importance relates mostly to how good a fit the candidate tenant is for the incubator. Newcos that are relatively unformed can develop and thrive in the right environment.

The definition of success of companies and incubators has recently been modified to include non-financial (community benefit, training) and financial benefits (economic impact) (Hewitt & Van Rensburg 2020; Laitinen and Chong 2006). In 1986, Smilor and Gill identified a number of properties needed for a successful incubator, which still seem relevant today (Lasrado et al. 2016) and are listed below:

- Onsite business expertise
- In-kind financial support
- Community support
- Entrepreneurial (financial) support
- Entrepreneurial network
- Entrepreneurial education
- Perception of success
- Selection process for tenants
- Connection to a university
- Concise milestones for progress

A 2022 study by the Brookings (Metro) Institution touched on the topic, "What makes a technology cluster a success?" I like this list because the paradigm includes a cultural component of success, as well as a financial one. It uses the term "cluster" but I would substitute "incubator" (paraphrased from Parilla et al. 2022a, b).

2.7 Defining Success for Life Sciences Incubator

Reasons for Incubator Success Beyond Tenant Selection: Borrowing reasons for an incubator success from Parilla (2022a) and Smilor and Gill (1986), we provide a modified list, adapting from our experience (for example, substituting an association with research organizations for association with a university), as described below:

- *Research and commercialization* refer to programs that help product development and fill in technical and business gaps the startup may have.
- *Entrepreneurship and capital access interventions* provide critical resources to young firms and entrepreneurs to support startup growth and innovation, typically through private equity facilitation, accelerator and incubator programs, and revolving loan funds.
- *Governance interventions* refer to resources related to helping with management, administration, capacity-building, and equity initiatives.
- *Association with a research organization* such as universities, research-oriented biopharma, or, in the case of the PABC, a nonprofit research institute can be a pipeline of new technologies and talent and, if colocalized within the incubator, can also provide a knowledge culture and, in the case of nonprofits, a sharing and collaborative environment with a range of services and amenities.
- *Talent development* such as K-12 programs, higher education nondegree and degree programs, workforce training programs, apprenticeships, internships, and other talent pipeline initiatives.

These qualities are generally consistent with the elements for incubator success that Smilor described in 1986. Added to their list are the additional parameters of community support and affiliations with a university, which can be a substituted for a research institution. If those are the reasons that an incubator and startup are successful, it is also important to define the goals and definition of success. I define the success of these enterprises in two complementary ways: (i) the success of the incubator and (ii) the success of its constituents (the startups and other entities depending on it).

Definition and metrics of incubator success: Apart from determining what makes an incubator successful, it is useful to define the metrics and qualities used to measure and quantify success. Borrowing from others (Al-Mubaraki and Busler 2013; Al-Mubaraki et al. 2013; Fernández et al. 2018; Hausberg and Korreck 2020; Lasrado et al. 2016) and based on our own experience, we propose the following criteria to *measure* the success of an incubator:

- *Success of Its Startups:* At the end of the day, the key definition of an incubator's success is how successful the entities are that have been nurtured in this environment.
- *Beneficial Economics:* The number of jobs created and retained, the amount of money spent by the incubator and its constituents as a direct result of its operations, and its secondary effects are all measurable and arguably the most important benefits and returns on investment to publicly funded incubators.

- *Cultural Contributions to the Community:* Although difficult to quantify, the cultural benefits of an incubator to its community range from the tangible (creating and maintaining attractive properties and buildings) to the less tangible, in which intellectual vigor and opportunities are made available (events, perceived opportunity access, a resource for displaced workers).
- *Beneficial Innovations:* New medicines, therapeutics, devices, and new information (as manifested in publications and patents) generated by the constituent entities and the people working at the incubator are certainly valuable products. These need not be commercial blockbuster successes, or even commercial successes at all, to be impactful. Even if a new innovation fails, if properly reported, it can be of informational benefit to society. Therapeutics, diagnostics, and medical devices entering clinical development and, of course, that end up being used (reach commercialization, for example) are measurable and represent the core innovation contribution of an incubator.
- *Numbers and Success of Trainees:* This metric includes the number of people trained, as well as their subsequent positions and accomplishments. Workforce training for incumbent and displaced workers is important. Academic, research, and entrepreneurial experiences for high school, college, graduate, medical, and postdoctoral trainees are also significant contribution.

Quantifying the Success of a Startup in the Incubator: The success of startups in the incubators is a function of the following factors, and the extent to which the business incubator can influence these factors will determine the success of the incubator. Factors by which the success of a startup can be judged include the following:

- *Funds Raised for Development:* This includes nondilutive and dilutive funding. Nondilutive funding investments are made without the expectation that equity in the Newco will be given in exchange for the investment. Dilutive funding investments are made with the expectation that an amount of equity in the Newco is given in exchange for the investment. Nondilutive funding can include grants from the U.S. Small Business programs such as Small Business Innovation Research (SBIR) and Small Business Technology Transfer (STTR) programs. Dilutive funding can be from early angels and later VCs, and sponsored contracts from for-profit and nonprofit organizations. SBIR and STTR funding is from the federal government, and in the life sciences, it's usually NIH funding to the startups. The Newco proposes its technology as a research project, initially as a feasibility study (Phase I), which in 2023 was limited (unless a special allowance is made) to $400,000 and 1 year. If successful, this can be followed by a Phase II development project and these can run into the low millions of dollars and extend for 2–3 years.

    The SBIR and STTR grants carry with them a value that goes beyond the direct dollar amounts. In addition to the obvious financial value of the nondilutive support for the Newco, the awards are made following a rigorous review of the proposal by experts in the field. So an SBIR or STTR award provides an enormous vote of confidence in the work from the expert scientific, medical community and strengthens the Newco's case in seeking other, dilutive investments. It can be a significant predictor of a startup's success.

Dilutive funding comes from investors such as angels who believe in the Newco, Angel groups (networks of individuals who organize funding the Newco as a group), and venture funds (formal corporations that have raised funds, themselves, with the intention of investing in Newcos). Finally, there are other companies that may fund the Newcos by making their own investments, providing sponsored research support, or providing funds in exchange for licensing or accessing a technology that the Newco can offer.

Company Sustainability: The number of years that a company continues to operate is certainly one metric of the success of a startup. However, companies that have quick exits as acquisitions should also be considered successful outcome.

Successful Innovation Development: The introduction of a new and useful technology, even if it falls short of commercialization or commercial success, can make a startup a social success, even if it is not a commercial success, and be a credit to the incubator.

Commercialization and Commercial Revenue: Products (medicines, reagents, diagnostics, devices, processes, and assays) brought to use and to market are clear measurable elements of success. Revenue from product sales or professional services is quantifiable. The profitability of Newcos is also a success indicator when it becomes revenue generating. Operating surpluses are the metric used to assess the financial and operational success of a nonprofit incubator.

Successful Exits: Although not unheard of, startups usually do not reach profitability or even commercialization and sales while in the incubators. The more typical milestone of success for a startup at an incubator is to deliver a successful exit for its investors by being acquired (for an amount that delivers a significant return to the investors and creators) or progress to an IPO or a reverse merger with a publicly traded company. IPOs are where the Newco are permitted (by financial regulators) to sell their shares (stock) to the public. It is a complicated and expensive process, and although there are several stock exchanges, they are all heavily regulated by federal securities authorities. The most traditional exchange for Newcos in biotechnology is the NASDAQ in New York City. Once permission is received from the regulators, the Newco can raise millions or hundreds of millions of dollars. The investors in the Newco can then sell their shares and experience a cash exit.

The reverse IPO or reverse merger is another way in which a Newco can sell shares of itself in the public capital market and alleviate the heavy financial costs of filing for an IPO on its own. In a reverse merger or reverse IPO the Newco, which is a private company but presumably has an attractive product or technology, is typically merged with a "zombie" company that is listed on the stock exchange. The zombie company (a needlessly pejorative description) is usually one that has a failed product, and its only value is their seat on the stock exchange. By merging with the zombie company, the Newco can now raise money through a public offering. It will be an IPO for the Newco, although the zombie company will have already had a formal IPO. Nevertheless, it is a very powerful fundraising tool and is a means to return cash to investors in the Newco.

2.8 Why Do Startups Fail?

According to the SBA and several other sources (Nobel 2011; Kusumaningtyas et al. 2021; Hewit 2020), about 31% of new businesses fail within two years, and between 50% and 75% fail after five years. Although the percentages vary depending on how the startup is defined, ventured-backed ones fare worse than the average. (Gage 2012). Perhaps this is because compared with startups overall, venture-backed companies are likely to be among the riskier. In theory, however, they offer greater rewards. Venture-backed companies are those that are most likely to be in a biotechnology business incubator, so this is the standard to which business incubators should be compared. By any measure, therefore, incubator-launched companies outperform their peers. Companies that launched in incubators remained in business five years later, a startling 87% of the time. Even considering how the percentages vary with the study, this number is generally consistent with the numbers we have seen in our incubators.

Why do some startups succeed and others fail? There has been some investigation into this question. Much fuss has been made about the "geodemographics" of startups. Meaning, being the right person in the right place. For example, California and Massachusetts are usually regarded as the best places to be and have the best demographics. However, the main reasons for success certainly involve much more than can be explained by geography and demographics. Below are the top reasons for failure, according to Kusumaningtyas et al. (2021) and Small Business Trends, although to be fair, some of the reasons given are more symptoms of failure rather than cause of failure:

- Ran out of cash: 38%
- No market need: 35%
- Got outnumbered (competition): 20%
- Flawed business model and poor business plan: 19%
- Regulatory challenges: 18%
- Pricing issues: 15%
- Not the right team: 14–23%
- Product mistimed: 10%
- Poor product: 8%
- Disharmony among investors: 7%
- Pivot gone bad (changed strategic direction): 6%
- Burned out: 5%

One thing missing from that list is luck. Bad luck must also play a role in business failure, but how to quantify this is unclear and subjective. Its impact, however, is real. For example, according to one source, "Opening a restaurant in March 2020 (at the beginning of the COVID-19 pandemic and public health closing of restaurants) would be bad timing and bad luck." (Delfino et al. 2023).

An acronym used to summarize failure characteristics is "SHELL," which is intended mostly for information technology but also seems applicable to life sciences and is consistent with the survey results reported above (Kusumaningtyas et al. 2021). Paraphrasing from this publication:

SHELL

Software is wrong: The business model is inconsistent with product plans
Hardware is wrong: product is wrong
Environment is wrong
Leadership: bad organization, incompetent management
Lifeware: no market, running out of cash.

How much of SHELL can an incubator influence? How much of SHELL goes into the attractiveness or appropriateness of the components of an incubator? Obviously, incubators have a great influence on the environment in which the startup operates. For a high-functioning incubator, almost all of the variables of SHELL can be supported. On the other hand, each element of SHELL can be a part of selection for entry into an incubator.

Returning to one of our original questions: *Have incubators delivered? Have they been of benefit to startups and economic development?* There is some debate as to the value of business incubators to the overall economy or their cost effectiveness, but for the most part, there is general agreement that business incubators are beneficial to the companies they incubate. Multiple studies, as summarized by Voisey et al. (2006), Hewit (2020), and Rathore and Agrawal (2020) agree that business incubators contribute significantly to the success of their participants/tenant companies. The measurable benefits were financial growth, patents obtained, and alliances established that would presumably not have occurred or would have occurred less frequently in less amount than if they were not in the incubator program.

The InBIA estimates that the 1100–1400 incubators in the U.S. have assisted nearly 50,000 companies that have provided about 200,000 full-time jobs and generated $15 billion in revenue annually.

According to the Kauffman Foundation report (West and Sundaramurthy 2019), the highest startup activity, overall, in their 2017 Index was California, Texas, Florida, Arizona, and Colorado (See Fig. 2.4). Smaller states with the greatest number of startups were Nevada, Oklahoma, Wyoming, Montana, and Idaho. These rankings largely parallel the number of incubators per state. Circumstantial evidence to be sure but consistent with the notion that incubators are a helpful, healthy part of the innovation economy. This provides a degree of circumstantial evidence that incubators are either partly responsible for or at least correlative with startups and innovation activity in a state.

References

Al-Mubaraki H, Busler M (2013) Business incubation as an economic development strategy: a literature review. Int J Manag 30(1):362–372

Al-Mubaraki HM, Busler M, Aruna M (2013) Toward a new vision for sustainability of incubator best practices model in the years to come. J Econ Sustain Dev 4(1):114–128

Azoulay P, Jones BF, Kim JD, Miranda J (2020) Age and high-growth entrepreneurship. Am Econ Rev Insights 2(1):65–82

Battelle Memorial Institute. (2007). Characteristics and trends in north American research parks. Columbus: Battelle. At https://www.aurp.net/index.php?Itemid=92&id=46&option=com_content&view=article

Bioworld Vitaltransformation (2023) https://vitaltransformation.com/2022/12/the-us-ecosystem-for-medicines-how-new-drug-innovations-get-to-patients/

Bone J, Allen O, Haley C (2017) Business incubators and accelerators: the national picture. BEIS Res Paper 7(1):11–14

Breitzman A (2008) Analysis of patent referencing to IEEE papers, conferences, and standards 1997–2007

Breitzman A, Hicks D (2008) An analysis of small business patents by industry and firm size

Brieger SA, Bäro A, Criaco G, Terjesen SA (2021) Entrepreneurs' age, institutions, and social value creation goals: a multicountry study. Small Bus Econ 57:425–453

Delfino D, Shepard D, Martinez-White X (2023) The percentage of businesses that fail and how to boost your chances of success. Lending Tree. https://www.lendingtree.com/business/small/failure-rate/

Eckhardt JT, Shane SA (2011) Industry changes in technology and complementary assets and the creation of high-growth firms. J Bus Ventur 26(4):412–430

Fairlie RW (2014) Kauffman index of entrepreneurial activity, 1996–2013. Available at SSRN 2424834.

Fairlie RW, Desai S (2020) 2019 Early-stage entrepreneurship the United States: National and State Report. Available at SSRN 3607936

Fernández ND, Arruti A, Markuerkiaga L, Nerea S (2018) The entrepreneurial university: a selection of good practices. J Entrep Educ 21:1–17. Firstsiteguide: https://firstsiteguide.com/startup-stats/

Gage D (2012) The venture capital secret: 3 out of 4 startups fail. September 19, 2012, The Wall Street Journal (Small Business) online.wsj.com/article/SB10000872396390443720204578004980476429190.html?mod=WSJ_business_LeftSecond_Highlights

Giglio P, Micklus A (2021) Biopharma dealmaking in 2020. Nat Rev Drug Discov 20(2):95–97

Global Entrepreneurship and Development Institute (2019) Global entrepreneurship index. Washington, DC. http://thegedi.org/global-entrepreneurship-and-development-index/

Grabow J (2023) Ernst and Young Q2 2023 venture capital investment trends. Available at https://www.ey.com/en_us/growth/venture-capital/q2-2023-venture-capital-investment-trends

Hackett SM, Dilts DM (2004) A systematic review of business incubation research. J Technol Transf 29(1):55–82

Hackett SM, Dilts DM (2008) Inside the black box of business incubation: study B – scale assessment, model refinement, and incubation outcomes. J Technol Transfer 33(5):439–471

Hassan NA (2020) University business incubators as a tool for accelerating entrepreneurship: theoretical perspective. Review of Economics and Political Science, 2020 May 20

Hausberg JP, Korreck S (2020) Business incubators and accelerators: a cocitation analysis-based, systematic literature review. J Technol Transf 45:151–176

Haynes KE, Qian H, Turner SC (2012a) The location of business support programs: does the knowledge context matter?. Entrepreneurship, Social Capital and Governance: Directions for the Sustainable Development and Competitiveness of Regions, p 302

References

Haynes KE, Qian H, Turner SC (2012b) Geographical dimensions of federal investment in small business development. Studies in applied geography and spatial analysis. Addressing real world issues, pp 144–159

Hewit S (2020) The Six key reasons why companies fail and how to plan for success. https://www.linkedin.com/pulse/6-key-reasons-why-remote-working-fails-how-plan-success-scott-hewitt/

Hewitt LM, Van Rensburg LJJ (2020) The role of business incubators in creating sustainable small and medium enterprises. South Afr J Entrep Small Bus Manag 12(1):9

Hicks D, Breitzman A Sr, Hamilton K, Narin F (2000) Research excellence and patented innovation. Sci Public Policy 27(5):310–320

Hicks D, Breitzman T, Olivastro D, Hamilton K (2001) The changing composition of innovative activity in the US—a portrait based on patent analysis. Res Policy 30(4):681–703

Hwang TJ, Franklin JM, Chen CT, Lauffenburger JC, Gyawali B, Kesselheim AS, Darrow JJ (2018) Efficacy, safety, and regulatory approval of Food and Drug Administration-designated breakthrough and nonbreakthrough cancer medicines. J Clin Oncol 36(18):1805–1812

Kennedy B, Tyson A, Funk C (2022) Pew foundation survey on how Americans perceive science. https://www.pewresearch.org/science/2022/02/15/trust-in-scientists-declines-appendix/

Kesselheim AS, Tan YT, Avorn J (2015a) The roles of academia, rare diseases, and repurposing in the development of the most transformative drugs. Health Aff 34(2):286–293

Kesselheim AS, Wang B, Franklin JM, Darrow JJ (2015b) Trends in utilization of FDA expedited drug development and approval programs, 1987–2014: cohort study. BMJ 351:h4633

Kneller R (2010) The importance of new companies for drug discovery: origins of a decade of new drugs. Nat Rev Drug Discov 9(11):867–882

Kusumaningtyas A, Bolo E, Chua S, Wiratama M, Tirdasari NL (2021) Why start-ups fail: cases, challenges, and solutions. In: Conference Toward ASEAN Chairmanship 2023 (TAC 23 2021). Atlantis Press, pp 155–159

Laitinen EK, Chong G (2006) How do small companies measure their performance. Probl Perspect Manag 4(3):49–68

Lasrado V, Sivo S, Ford C, O'Neal T, Garibay I (2016) Do graduated university incubator firms benefit from their relationship with university incubators? J Technol Transf 41(2):205–219

Micklus A, Giglio P (2020) Biopharma dealmaking in 2019. Nat Rev Drug Discov 19(2):87–89

Munos B (2009) Lessons from 60 years of pharmaceutical innovation. Nat Rev Drug Discov 8:959–968. https://doi.org/10.1038/nrd2961

Nicholls-Nixon CL, Valliere D, Gedeon SA, Wise S (2021) Entrepreneurial ecosystems and the lifecycle of university business incubators: an integrative case study. Int Entrep Manag J 17(2):809–837

Nobel C (2011) Why companies fail–and how their founders can bounce back. Harvard Business School, Boston

O'Loughlin G, Bowen H, Schulthess (2023) The US ecosystem for medicines (2011–2020). https://vitaltransformation.com/wp-content/uploads/2022/12/Where-do-new-medicines-originate_FINAL2022_12_05.pdf

Parilla J, Haskins G, Muro M (2022a) The future of place based economic policy. https://www.brookings.edu/wp-content/uploads/2022/11/EDA-BBBRC_final.pdf

Parilla J, Donhahue R, Martinez S (2022b) Institutionalizing inclusive growth: rewiring systems to rebuild local economies. Brookings Institute. https://www.brookings.edu/articles/institutionalizing-inclusive-growth-rewiring-systems-to-rebuild-local-economies/

Pitchbook (2023) NVCA venture monitor. https://nvca.org/wp-content/uploads/2022/01/img4.png

Rathore RS, Agrawal R (2020) Measuring performance of business incubators: a literature review and theoretical framework development. In e-journal-First Pan IIT International Management Conference–2018

Redit C (2022) Pharma backs off biotech acquisitions. Nat Biotechnol 40:1546–1550. https://doi.org/10.1038/d41573-023-00012-0

Shane SA (2008) The illusions of entrepreneurship: the costly myths that entrepreneurs, investors, and policy makers live by. Yale University Press

Small Business Trends (2020) (web site) https://smallbiztrends.com/2023/07/startup-statistics.html
Smilor RW, Gill MD (1986) The new business incubator: linking talent, technology, capital, and know-how. (No Title)
Start up ranking.com. https://www.startupranking.com/countries
Statistica (2023) https://www.statista.com/ and https://www.statista.com/topics/4733/startups-worldwide/#topicOverview
Tareque I et al (2017) Kaufman foundation start up index 2017 report
Tsaplin E, Pozdeeva Y (2017) International strategies of business incubation: the USA, Germany and Russia. Int J Innov 5(1):32–45.
U.S. Small Business Administration Small Business Profile (2020) Office of advocacy. https://advocacy.sba.gov/2020/05/20/2020-small-business-profiles-for-the-states-and-territories/
Voisey P, Gornall L, Jones P, Thomas B (2006) The measurement of success in a business incubation project. J Small Bus Enterp Dev 13(3):454–468
West C, Sundaramurthy G (2019) The successful college drop out is rare. The majority of start up executives have advanced degrees. https://www.kauffmanfellows.org/journal/startup-degrees
Wiggins J, Gibson DV (2003) Overview of US incubators and the case of the Austin Technology Incubator. Int J Entrep Innov Manag 3(1-2):56–66

Chapter 3
Pennsylvania Biotechnology Center: A How-to Manual

As mentioned in the previous chapters, there are multiple models of business and life science incubators. Those created and run by universities are the most common, such as the University City Sciences Center in Philadelphia. There are some very successful for-profit incubators, such as those from J Labs, SOS, Deerfield, and Alexandria. Governmental or municipal-run incubators exist but are less common. Incubators run by nonprofit, nongovernmental organizations, such as the Pennsylvania Biotechnology Center (PABC), are in the minority.

We will explore the PABC in detail because of its success and because of our familiarity with it. A deep dive into the PABC will provide specifics about an incubator operation, funding, and strategic planning. This part of the book may be useful as a "how-to" manual for those interested in biotechnology incubators, in general, and for those who may actually be involved in the operation of incubators.

The PABC is one of the most successful life sciences incubators in the U.S. Both Kolbetree (2021) and InBIA (2017) rank the PABC in their top 10 based on a variety of factors, including economic impact, Initial Public Offerings (IPOs) and funds raised. The PABC has been achieving these milestones of success from almost its beginning through the present, which is uncommon even for a mission-driven nonprofit organization. This was discussed previously but will be revisited below. Much about the PABC is different from both for-profit and other nonprofit incubators. These differences may explain some of its exceptional success and ability to thrive even in very difficult economic times. For example, between 2007 and 2009 and in 2020, the only times in the past 50 years when startup numbers in the U.S. dipped, startup creation at the PABC thrived.

The PABC entrepreneurs are slightly older and more experienced than the average entrepreneur. A key finding from Albort-Morant and Oghazi (2016) is that the entrepreneurs who "find support from incubators most useful ... are young, well educated [college or graduate education], have professional experience, and come from a family with a history of entrepreneurship." If the average age of a startup founder in the U.S. is someone in their late 30s and early 40s (Azoulay et al. 2020; Brieger et al. 2021), at the PABC, it is closer to the early 50s.

That said, the PABC certainly has its share of young entrepreneurs, though I hasten to add that the most successful entrepreneurs at the PABC have been mid- to even late-career scientists from large pharmaceuticals and senior faculty from major universities. Many entrepreneurs at the PABC also come from nonprofit

© The Author(s), under exclusive license to Springer Nature Switzerland AG 2024
T. M. Block, *Curing Disease from the Ground Up*,
https://doi.org/10.1007/978-3-031-56148-1_3

organizations based at PABC, such as the Baruch S. Blumberg Institute (Blumberg Institute/ BSBI). This is because the Blumberg Institute is, by definition, a translational research organization that is constantly creating new technologies. None of these characteristics fit the profile described by Albort-Morant and Oghazi (2016). This may help explain the unique contribution of the PABC and the key to its paradoxical success, even during periods of economic distress. The entrepreneurs who have been successful at the PABC reflect the greater pharma community in our region, which is something of a bedroom community for many pharma and biotech companies. They actually dramatize how there can be exceptions to the standard profile of a successful entrepreneur. In fact, the strength of the PABC, and in many ways, our distinction, is how the typical entrepreneur is more mature relative to our affiliate location, B Labs in Philadelphia.

3.1 Description of the Campus and Its Physical Layout

To obtain a full picture of the PABC as an example of a business incubator that began in an old warehouse and grew concentrically, it is useful to know its physical footprint. Moreover, for anyone interested in business incubator operations, knowing the property layout, floor plan, facility details, and tenant mix may also be helpful. I certainly benefited from this type of information when we started in 2006.

The first building for PABC was a 60,000 ft^2 mail distribution warehouse that was renovated into an office building and research labs. By 2022, the PABC was composed of four interconnected buildings totaling ~150,000 ft^2 on a 14.5-acre campus. This included a new 38,000 ft^2 building that opened in the Spring of 2022. An overview of the campus is shown in Fig. 3.1, with a floor plan. With the completion of the new building in 2022, there are smart, new state-of-the-art labs, offices, meeting rooms, and common services for the Center, and the specifics are provided in the Table.

As of 2023, the PABC has more than 400 people working in more than 50 different organizations on its campus. PABC calls tenants "stakeholders", "residents" or "members" to emphasize the community nature of their tenancy. The stakeholders range from part-time, one-person businesses to a 50-student high school program and the nonprofits, the Hepatitis B Foundation (HBF) and the Blumberg Institute, with more than 100 scientists, public health experts and staff. It is, therefore, necessary to have a range of work spaces, ranging from very small labs to multicompany shared labs. Larger multiroom labs for single companies and standard single academic labs, classrooms, offices, and meeting spaces are all necessary. The allocations of space are summarized in Table 3.1.

The stakeholders occupying the most space and longest-term lease holders at the PABC are the HBF and the Blumberg Institute, which occupy ~19,000 ft^2 comprising 13 labs, 25 offices, and meeting rooms. As discussed before, the Blumberg Institute is a nonprofit translational research organization. This is important since the scientists working there are regularly creating new intellectual property and are

3.2 Power, Water, Information Technology (IT), and Infrastructure Needs... 85

Panoramic of the campus (A). Floor plan (B) and profile of the Baruch S. Blumberg Institute and accelerator wing (C)

Fig. 3.1 Panoramic of the campus (**a**). Floor plan (**b**) and profile of the Baruch S. Blumberg Institute and accelerator wing (**c**)

Table 3.1 Occupant/tenant profile at the PABC

Type	Number	Ft² each	Ft² occupied	No staff/employees
Large company	1	>19,000	30,000	30
HBF/BSBI:	1	19,000	19,000	100
Mid-sized companies	7	3000–19,000	23,000	70
Smaller mid-sized companies	7	1000–3000	12,000	40
Small microsized companies	22	<1000	6000	30
Academics	Education		3000	50 students, 3 employees

themselves (or by technology transfer) creating new startups. These activities ensure that the PABC has reliable, responsible, and programmatically productive stakeholders. The HBF and the Blumberg Institute rent space, even if and when outside demand for space might dip. They provide a stable, steady population of scientists willing to collaborate with almost all stakeholders from the startups and are willing to share resources and equipment, helping maintain a collegial environment. Currently, the tenant distribution ranges from sizes that rival those of the Blumberg Institute to companies with only one employee occupying a desk space. See Table 3.2.

3.2 Power, Water, Information Technology (IT), and Infrastructure Needs of the Incubator

It's one thing to construct a large building on a plot of land and give it the glamorous designation of a biotechnology center. Easy. It's another thing to power and fit it out so that the Center runs smoothly. We learned early on that a technology incubator, with expensive and energy-intensive equipment, has many facility needs that are

Table 3.2 Space distribution of the PABC

Unit type	Unit size (ft^2)	# of units
Lab	150–350	23
Lab	350–750	37
Lab	2,400	1
Cluster shared space offices	1200	3
Shell	5,000	1
Walk in −20	1,400	1
Cold rooms	250	2
Common equipment	100–200	3
Warehouse/manufacturing	34,000	1
Event room	For a 70 person room	1
	For a 200 person	1
Conf rooms	For a 10 person	5
Conf rooms	For a 15–20 person	3
Offices	For a 120–150 person	100

uncompromising and distinguish it from most other commercial operations. Water usage and waste disposal, uninterruptable and steady high-quality power, emergency power and service redundancies, and security support must all meet industry standards and regulations. Again, we learned this after we opened, and yes, it was as unnerving as learning how to fly a plane after you are airborne and alone.

PABC is not in a major city with plentiful and redundant services. We aren't even in the center of the small town of Doylestown. The Center is on what was an infrequently trafficked country road, a couple of miles from the town center. Many of the PABC services had to be designed, recreated, or adapted. This section of the book, therefore, describes the current services and infrastructure at the PABC to provide details to anyone interested in what is required to operate a ~ 150,000 ft^2 biotechnology incubator outside of the power grid of a major city.

The original building was built around 1970 and was a two story warehouse with open office spaces for cubicles. After major renovations were made for the PABC, most of the building was retrofitted with individual offices, meeting spaces, and research labs that also included a small NMR lab in operation in the building. There is "normal" power for the labs and offices and "emergency" power provided by 300 kW and 150 kW diesel-powered emergency generators. Automatic transfer switches move power between normal and emergency in the event of a power outage. Most of the lab refrigerators and freezers, exhaust fans, and other critical equipment are on emergency power. The offices and meeting rooms are primarily serviced by various packaged roof top units (RTUs) to supply conditioned air, either cooled or heated, and ducted throughout the office areas.

The HVAC in this building is a four-pipe hot/cold/return circulating water system with three boilers for heating and a 75 ton air-cooled water chiller. Hot or cold water is circulated to individual office wall units for offices on the perimeter of the building and uses air-handling fan units for common spaces and offices/cubicle areas in the core areas of the building.

Labs have a once through make-up air system (MUA) where air is ducted to a fan coil unit for each lab and either heated (electric heat) or cooled (chilled water) to a set temperature. There are four Aaon MUA units on the roof, supplying air to the labs. Chilled water is provided by two rooftop-mounted air-cooled 120 ton chillers and circulated to the main MUA and to each fan coil unit. The labs also each have an exhaust fan to remove fumes from the chemical fume hoods and to create a negative pressure in each lab to prevent the spread of any contamination.

Some of the equipment in the building is on a building automation system (BAS) to control the settings and temperatures of the labs and offices. Newer building designs make use of additional controls to reduce air flows through the labs when they are not occupied or when the hood sash is closed to save significant energy.

For our new building, completed in 2022, we were able to custom design it with more modern power, services, and other amenities. This is also a two-story building, with labs and offices on each floor, a new administration suite and visitors lobby, and a larger conference room. There is also some undeveloped shell space on the second floor to be developed into additional labs and offices as funding and needs are established. Some of the labs in this building—all biological, not chemical—include a small-tissue culture room within the lab footprint, and some labs were designed as double labs to meet tenant needs.

The HVAC in this new building also uses packaged air RTUs for the offices but uses a manifolded MUA and exhaust system for the lab areas. Currently, two Aaon DX cooled units and two vertical exhaust fans are in place, but the system was designed to be able to add another Aaon and exhaust to the manifold for future development. A full BAS building automation system is in place for the new systems and will be tied in with other buildings for better overall HVAC control.

New power was run from the street for this building, along with a 750 kW diesel generator. The entire building is on emergency power; however, some reduction in the MUA may occur if the building is under high electrical load. New water services were installed from the street so that all buildings were supplied by city water rather than the original well. The well is still in use to feed a fire tank storing water for a diesel fire pump, which serves the warehouse area. The remainder of the buildings are served by a new electric fire pump with water supplied from the street. For IT, high speed internet connectivity, data storage and management systems, rooms with adequate power and cooling for servers, cybersecurity systems and professional support for these services are now essential.

3.3 Management of the PABC

The way in which the PABC is managed is unusual compared with other incubators, which are mostly operated by academic and nongovernmental organizations or for-profit developers (Clark et al. 2018; de Bem et al. 2020; Mubarak Al-Mubaraki and

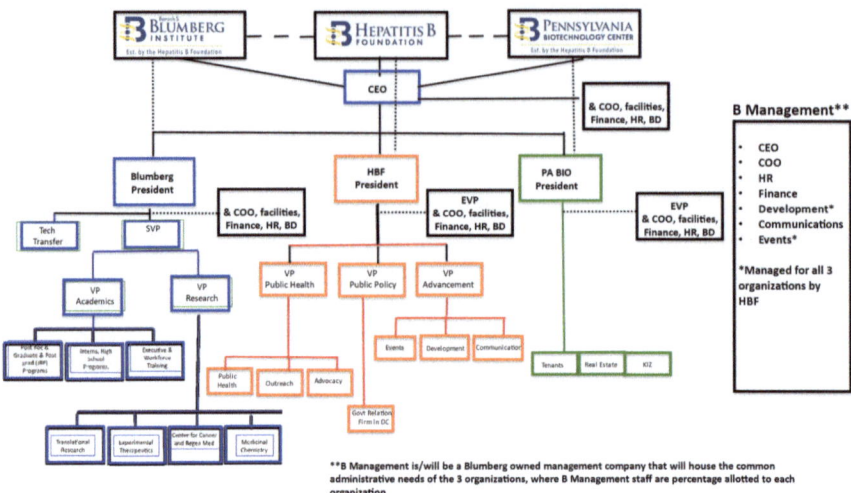

Fig. 3.2 Organizational chart of BSBI, HBF and PABC

Busler 2010). That said, the PABC management model provides an example of how companion organizations can cooperate to produce greater resources for the incubator than would otherwise be possible. In the case of the PABC and its companion organizations, all three are nonprofits. As a nonprofit 501c3 organization, the PABC is governed by a Board of Directors (5–7 people) who are appointed by the Board of the Hepatitis B Foundation, which itself has 21 members, and the Blumberg Institute is governed by up to 9 Board Directors from the HBF. The HBF is the sole "member" of the PABC nonprofit corporation. An overall organizational chart of all three entities is shown in Fig. 3.2.

As shown in Fig. 3.2, there is a president of the PABC who is responsible to the CEO and the Board of the HBF. When the PABC was established, I served as both the president of the PABC and CEO of the three affiliated nonprofits. This arrangement remains true in 2023, but with a paid CEO. Although it is not a mandate of the bylaws, having a CEO who manages all three non profits has assured smooth integration and coordination of the three organizations. Given the central role the Blumberg Institute serves in managing and populating the PABC, this coordination is especially useful.

Perhaps surprisingly, the PABC has no employees. It is managed and operated by employees of the Blumberg Institute, with which it has a management contract agreement. In this way, there can be an efficient, nonduplicative, consolidated, and coordinated operations management of the PABC, Blumberg Institute, and HBF. The PABC, which has relatively modest revenues that would otherwise severely limit the size of its staff, can have an operations depth it would otherwise not be able to afford. Similarly, the Blumberg Institute and HBF can also have operations, IT, and facilities support they would struggle to have otherwise.

3.3 Management of the PABC

These limitations were particularly pronounced at the beginning of the PABC in 2006 but are still true in 2023. Even with an annual revenue stream of ~$5 million per year, the PABC would only be able to employ a small staff and struggle to fund executives and managers. This is also true for the Blumberg Institute, with a budget of ~$16 million per year, and the HBF at ~$4 million per year. The things these organizations can spend money on are often restricted by the funding source. Collectively, however, with shared interests and common, complementary needs, departments that provide many of the amenities and functions of a larger organization are possible: grant management, financial and HR services, IT, facilities and janitorial maintenance, fundraising and event planning, and other administrative functions are now all consolidated.

B Management

The Blumberg Institute created a for-profit corporation called B Management to provide for the professional management services offered to the entire PABC, Blumberg Institute, and HBF enterprise. B Management can also offer these management services to other organizations. It is 100% owned by the Blumberg Institute, making it a subsidiary but as of this writing, it is not fully executed, and functions as a division within the Blumberg Institute. Recently, to accommodate our partnerships and management of other business incubators, B Management has hired new employees who are fully integrated into our nonprofit culture.

B Management and the senior administration of the PABC, Blumberg Institute (BSBI), and HBF provide a stable core community for the entire PABC. These positions are mostly full-time, onsite, and are listed in Table 3.3. The effort that every position puts into each organization is also indicated and should be noted as an approximation. However, the role overlap between the different organizations highlights the complementary roles of each position. Many positions span all three organizations, thus allowing a single individual to provide service to multiple organizations. This means that any of the three institutions can have a senior position filled at less than 100% cost. The smooth integration and assurance that no single institution is left deficient in any area are ensured by both the enthusiasm of the individual and the CEO, who has responsibilities to all three companion nonprofit organizations.

B Management as a services organization for hire: The PABC was asked by many of the startups and other nonprofits at the Center to provide administrative services because they were comprised of fewer than 10 scientists and staff. Often, the startups had only one or two employees and needed help with a range of things, including leasing, IT, grant writing, business planning, facilities support, compliance, legal, board development, business development, personnel, and so on. Although some of these services were provided to tenants of the PABC as a part of their rent, there were many services performed by the PABC (via the Blumberg Institute management contract) that were not. Moreover, the PABC was asked to replicate its success at other physical locations. To achieve this, B Management, which was originally created to run the three nonprofits, was adapted to be available

Table 3.3 Staff managing the PABC, BSBI, and HBF

Title (# admin. assistants, and staff)	Title applies to			Organization and estimated percentage distribution of effort		
	PABC	BSBI	HBF	PABC	BSBI	HBF
Office of the CEO						
CEO	x	x	x	25	45	30
Pres PABC (1)	x			90	5	5
COO	x	x	x	40	40	20
CFO	x	x	x	25	45	30
Dir Accounting (2)	x	x	x	25	45	30
Vice President Dev (2)	x	x	x	25	45	30
Dir Operations, Leasing	x			90	5	5
Dir Operations B labs	x	x	x	45	45	10
Dir IT	x	x	x	50	25	25
IT & Audio manager	x	x	x	50	25	25
Dir HR	x	x	x	10	45	45
Dir Facilities	x	x	x	50	40	10
KIZ Coordinator	x			90	10	0
Dir Events (1)	x	x	x	33	33	33
Dir Communications (2)	x	x	x	30	20	50
Res Safety Officer	x	x	x	50	50	
House keeping	x			80	15	5
House keeping	x			80	15	5
Grounds[b]	x					
Reception	x			90	5	5
Reception	x			90	5	5
Reception	x			90	5	5
BSBI	PABC	BSBI	HBF	PABC	BSBI	HBF
President BSBI (1)		x		5	90	5
SVP Research & CSO		x		0	95	5
Dir Academics		x			80	20
Dir Med Chem		x			100	
Dir Exp Thera		x			100	
Dir Early Detect		x			100	
Dir Regen Med		x			100	
Dir Grants & Res Admin (1)		x		5	85	10
Legal general[a]		x		10	60	30
Legal IP[a]		x		10	85	5
Bus Dev		x		5	85	10
Other faculty, scientists and staff (65)					100	
HBF	PABC	BSBI	HBF	PABC	BSBI	HBF
President HBF (1)			x	0	10	90
Dir Pub Health			x		5	95

(continued)

Table 3.3 (continued)

Title (# admin. assistants, and staff)	Title applies to			Organization and estimated percentage distribution of effort		
	PABC	BSBI	HBF	PABC	BSBI	HBF
Dir Outreach			x		5	95
Dir Advocacy			x		5	95
Dir Social Media			x		5	95
Other professionals and staff (21)				100		

(x) Highlights indicates a "B management" service offered to Newcos at the PABC and B Labs in Philadelphia
[a]Office onsite, but part time
[b]By contract

on a fee basis to the startups at the PABC and other incubators. Its professional services were already being provided, fractionally, to each of the three nonprofits and, with the adaptations, could be offered in a "menu" to tenants and other organizations at the Center and elsewhere.

B Management is now contracted by for-profit companies and other nonprofit organizations at the PABC (and offsite), and each is billed accordingly. Current leadership at the PABC has a vision in which B Management becomes a multimillion-dollar services corporation that will be another stream of funds for the Blumberg Institute and HBF, and allow for economies of scale in recruiting expensive talent. In 2022, B Management brought in ~$1.5 million per year to the Blumberg Institute, of which $600,000 is from outside contracts.

3.4 The Ins and Outs of Being a Tenant Member

3.4.1 Choosing Tenants, Advertising

Getting tenants in and out of an incubator is one of the most important and touchy things management will face. Along these lines, the selection of the specific tenant startups to be admitted to the incubator is one of the most important factors for both of their successes, according to much of the literature on the subject (Hackett and Dilts 2008; Fern et al. 2018; Fernández et al. 2018). Although I am uncertain about agreeing with that level of priority, maintaining a culture of innovation often does require finding startups and entrepreneurs that are compatible with the PABC and are most likely to benefit from what we have to offer. This can mean actual talent scouting or, more often, judging proposals from entrepreneurs wishing to take space at the Center.

Some of what we look for is a bit self-serving. For example, and to justify our reasoning, in the beginning, the PABC and the Blumberg Institute lacked certain major equipment, such as high-end fluorescence-activated cell sorters (FACs), nuclear magnetic resonance (NMR) spectroscopy and high-end mass spectrometers. We have since acquired that kind of expensive equipment and can make it available to residents of the PABC, but initially we looked for startup entrepreneurs that could bring high-end equipment we didn't have. We then made arrangements with those entrepreneurs to make the equipment available to others at the Center, usually through fees for service and collaboration.

This approach worked not only for equipment but also for professional services. For example, because the Blumberg Institute was generating promising, but early, hits in our antiviral screens for HBV drugs, we wanted to have access to medicinal chemists who could improve compound design. We even sought scientists who could create custom chemistry companies—Michael Xu's Pharmabridge and Alan Reitz's Fox Chase Chemical Diversity Company. These companies collaborated with those at the Blumberg Institute and other scientists and companies at the PABC, giving us an industry standard medicinal chemistry capability certainly not seen at most universities.

The ability of a prospective tenant to provide service or complement others at the PABC was not the only or even highest priority in our consideration of them, but it was a factor. That said, bringing in a tenant with the ability to complement the work of others at the Center had the additional benefit of fostering cooperation and greater awareness of the new tenants. If a prospective tenant refused to cooperate (I think this happened once), we would likely demure or perhaps wonder about this as an indication of a lack of compatibility with our Center, which emphasizes cooperation. I can say that the willingness and enthusiasm of a prospective entrepreneur to embrace the PABC philosophy of cooperation has been a pretty good indication of how well they fit into the knowledge community and how well they will thrive in the community.

Initially, the PABC had an admissions committee to evaluate prospective tenants comprised of the CEO, a PABC and/or a Blumberg Institute board member, and representatives from the Blumberg Institute and the tenant startups. The decisions are now made by B Management leadership but follow the same or similar criteria, with the understanding that much of the criteria overlap and can be subjective.

Will the tenant benefit from being at the PABC? Although subjective, based on professional judgment, the determination of whether or not the Center will benefit from having a tenant is a major component of the overall assessment we use. Prospective tenants developing new antivirals, anticancer medicines, and devices or need access to our compound libraries, virological labs or medicinal chemistry expertise are more likely to benefit than prospects making cosmetics, for example. The other criteria, listed below, are largely subspecies of these first criteria.

Will the PABC benefit from having the tenant? Prospects with exciting projects, stimulating individuals with attractive demand skills, and, of course, major equipment, can bring valuable expertise, and, if they are willing to collaborate or at least cooperate with others at the Center, there will be tremendous synergy.

Thematic compatibility: Are their programs and goals enhancing, compatible, and/or complementary to the Center? For example, a company proposing a new antiviral is more compatible with us than a company proposing material science development.

Leadership compatibility: Entrepreneurs who bring wisdom and professionalism will be a big plus to the Center and are more likely to have successful startups. This is not to say the startup doesn't need or wouldn't benefit from complementary professional talent, which is where we can help. However, we try to be mindful of entrepreneurs who may be disruptive to themselves. The PABC is a lively, interactive culture that, in some respects, resembles an academic institution (with seminars, hallway discussions, sharing of reagents, and so on). Some entrepreneurs may find that degree of openness and frequency of interaction uncomfortable.

Resources the tenant needs or might bring: This criteria is similar to and consistent with the compatibility and mutual benefit considerations, as stated above, but is focused on the specific equipment the prospective tenant will bring.

Financial soundness of the tenant: The importance of this factor depends on the amount of space and resources the prospective entrepreneur will need and consume. There are greater risks with tenants needing multiple labs and use of our major equipment compared with those who need only a small amount of space and resources. Most prospective tenants are financially risky, and accepting this is pretty much baked into the PABC model.

Business, intellectual property, and financial promise of the tenant: We rank this concern lower than the other priorities. This is because at the time a tenant is being considered for entry, it is usually difficult for us to predict the likelihood that they will be a business success. If a prospect ranks high in the other categories, it is likely that even those with a less professional business plan will excel.

Benefits to the region: Some prospective startups may not herald enormous financial gains, but they may be more likely to establish themselves in our region or may be proposing products and/or projects that are particularly useful to the region. Some professional services and "information" companies, such as those proposing workforce training, legal and medical offices, and financial services, come to mind.

3.4.2 Moving Tenants In and Out

One important quality that may distinguish the PABC from many other incubators is that it does not require tenants leave after any particular period of tenancy. Many, if not most, incubators and accelerators have short, well-defined periods of time during which a tenant may remain. For some accelerators, this can be only a matter of months. For most incubators, the time is a year or at most a few years. For us at the PABC, how long a tenant stays is up to them. Understandably, those with suddenly exploding space needs beyond our capacity have to leave. However, for the most part, we encourage startups to grow within the Center as long as possible. Our reasoning for this is discussed below.

The life cycle of a startup at the PABC often means going from a single office to a lab bench in a shared lab with access to common equipment and services, to an independent lab, and then to multiple labs. When a startup at the PABC raises sufficient investment such that it requires thousands of square feet, we struggle to either find it on our grounds (usually my preference) or more typically and necessarily find something nearby. Having mature companies and retaining mature companies on our campus is a great resource for the new, small, often single-person startups that make up the majority of users at the PABC. Mature companies have people who can mentor junior companies. Experienced entrepreneurs at the PABC willing to work with the less experienced startups are an attractive quality to the early-stage tenants here. In addition, although not a priority, mature companies are usually more reliable payers of rent.

The average stay at the PABC for a Newco is probably more like 5–7 years, rather than 1–2 years at other incubators. The Newcos even stay after raising significant cash, through acquisition and IPOs. That said, to be honest, we note that several companies have moved significant operations out of the PABC after an IPO or major raise, but even then, they frequently leave behind some division of their business. In addition, successful entrepreneurs often seem to return to the PABC with completely new ventures after the companies they created have been acquired.

3.5 Tangible and Intangible Support to Startups at the PABC

The PABC, through the Blumberg Institute management contract and its employees, can provide valuable assistance and unique resources to startups and their employees. In addition, the PABC has developed investment vehicles. Both means of support are summarized below.

Shared common equipment: There is access to major equipment, including low-temperature commercial freezers, ultracentrifuges, liquid-handling robots, mass spectrometry, confocal microscopy, digital PCRs, etc. Through onsite companies, there is access to NMR and FACS sorters.

In-kind financial support: This can be subsidized rent, forbearance, and even rent forgivingness in some cases. The PABC also provides direct investments through our own programs and, of course, less formally through angel networks to whom we make introductions, as outlined below.

Cash investment opportunities: There are formal and informal opportunities for startups to meet angel investors and VCs, and the Blumberg Institute faculty frequently collaborate with startups and help write grants to state and federal programs (i.e., PA Ben Franklin Partners, BioAdvance, SBIR and STTR grants, NIH U-01 grants, etc.).

Academic Entrepreneurs and Lab Scholarship Program: This is a program in which technologies from universities (or from independent entrepreneurs who have left their pharma employer) are identified as of interest to the PABC and the Blumberg Institute, and the entrepreneurs are offered investments. Typically, a faculty member is offered funding to come to the PABC and start a Newco based on their technology and may take a sabbatical from their university to do this. Or more typically, a member of their lab (e.g., post-doctoral student) will come to the PABC, and the faculty member will be on Newco's Scientific Advisory Board (SAB). The investments are in exchange for equity. All necessary services are provided by CEOs, business planning, and SAB formation to these labs and personnel. Typical investments are between $250,000–300,000 per year (in 2022) for 2 years and/or subsidized labs. The sources of funds are grants from the Commonwealth of PA and discretionary support from the PABC and the Blumberg Institute, which allows them to bring very specific and deliberate talent and technologies to the Center that are considered complementary and enhancing to their programs.

Hatch Biofund: This is a more conventionally styled venture fund, with the qualifier that its mission is to invest in companies at the PABC and its affiliated B Labs joint venture incubator located in Philadelphia. Investments are in the 1–3 million dollar range with follow-up support possible. The rationale for the fund is that Hatch is able to keep a close eye on the Newcos at PABC and B Labs that have outperformed their peers, thus getting on the ground floor of investing. Sources of funds include private and institutional investors and a pharma company. Hatch Biofund is partly owned by the Blumberg Institute but independently managed and currently has $34 million to invest with a goal of reaching $50 million. If successful, the second and third funds are planned.

Keystone Innovation Zone: The PABC is one of the state's 29 Keystone Innovation Zones (KIZs). These are specific areas within the state designated by the legislature. In some cases, KIZs encompass many square miles, but in our case, it includes primarily our main 14.5 acre PABC campus and our original building at Delaware Valley College (now University). The main benefit of being in a KIZ is that qualified tenant companies may receive tradable tax credits. When KIZs were initially created, there were a number of other benefits (e.g., specific state grants for Newcos in the KIZs), but these are no longer offered. That said, as of 2023, the KIZ tax loss transfer program can be worth $100,000 per year to the Newcos, and there is always the possibility that grant programs will be restored.

Thriving knowledge community: The PABC is a lively knowledge community, and we think this is an important, attractive, and beneficial quality for entrepreneurs. The physical spaces encourage mingling and talking, with lots of meeting rooms and an Idea Cafe that offers free coffee and plenty of informal seating. We maintain a schedule of speakers ranging from outside experts of international renown to regular seminars and business presentations from internal speakers (e.g., a 'company spotlight' that features our own tenants, research rounds, journal clubs) and annual events for the tri-state region, such as the Entrepreneur Spotlight and Regional Biotech Conference. We also have workforce training and academic programs, including high school, college, Master's, Ph.D., and fellows working at the Blumberg Institute and, where appropriate, at the startups.

There are services for IT, biostatistics and data science support (e.g., the Blumberg Precision Medicine Center), as well as legal support ranging from intellectual property and business contracts to immigration and general law. Access to chemical libraries, ChemdDiv software, electronic journals and other resources not easily attainable by startups. Through B Management and an onsite for-profit company Artemis Solutions, there is assistance in business planning, grant writing, grant submissions, bookkeeping, payroll and other administrative functions. We also have onsite CROs and consultants that can help find and provide transitional executive management.

Joint appointments: When appropriate, the Blumberg Institute may offer academic titles to startup entrepreneurs, and in some cases, these appointments can include stipends. The opportunities range from recent college graduates to senior scientists. For college graduates, we have a program called Junior Research Fellows or JRFs. We also have postdoctoral fellows working at the Center. Both JRFs and postdoctoral fellows can be assigned to startup labs, and thus, the trainee benefits from the semicommercial experience and the startup benefits from the technical help. We may offer professor-level titles in which the startup entrepreneur receives payment from the Blumberg Institute to provide teaching or other service work for its faculty. These arrangements can foster community, collaboration, and mutual awareness among stakeholders within the Center's community (often resulting in joint grant submissions, memberships on SABs, etc.), as well as, financial support for the entrepreneur.

Management services provided to select (willing) startup companies: The PABC and the Blumberg Institute make their technology development infrastructure available to the entrepreneurs at the Center. They have, as a matter of self-need, developed a punch list and mechanism for identifying and evaluating new technologies and developing them into new companies. The need came about because of the continuous pipeline of self-generated Newcos from discoveries made at the Blumberg Institute. There is a steady supply of spinout companies from the Blumberg Institute, which is, by definition, an academic translational research organization. Its scientists are dedicated to the discovery of new diagnostics and therapeutics to improve the lives of those affected by HBV and liver cancer. They, therefore, produce a steady stream of innovations that are intended to be products for human use. Moreover, this is often done with colleagues from other nonprofit institutions and universities, bringing even more technologies that have development potential to the attention of the PABC leadership. The PABC and the Blumberg Institute have worked together to develop processes to identify and develop promising technology that is determined to be a good fit for the Center. These protocols now provide road maps to create Newcos that develop the technologies. Mechanisms to fund the Newcos from seed to early stage are also routine for the PABC and the Blumberg Institute.

The approaches we use are outlined below in part as a punch list of services and tasks. These services, which prepare the Newco for favorable consideration of financial support from our funds and other investors, are made available to any willing and qualified Newco in the PABC and its partner incubators. We are now extending our services to universities, research institutes, and hospital systems. They can opt into the services we describe at any step.

3.5 Tangible and Intangible Support to Startups at the PABC

Table 3.4 Recommended "Punch List" of actions to be taken to launch each "Newco"/Portfolio Company (managed by the Portfolio Review Committee)[a]

Action needed	Time (within weeks of Arrival)	Goal
Technology Coordinator[b]	Time of invention disclosure	Technology and or Newco concept reported to the Technology Coordinator who sets process in motion
Project Manager[c]	Instant	Assures actions below are occurring in a timely way
Portfolio Oversight[c]	Instant	Assures Project management and business and progress is occurring
Incorporation[d]	Instant	LLC, C corp, other, filings
Initial Executives[e]	Instant	From the PABC network or outside consultants
Fund raising[e]	4 weeks	Fund raises, business plans, development plans
Boards formation[e]	4 weeks	BoD, SABs
Administrative[f]: Payroll	Instant	Self explanatory
Benefits	Instant	Self explanatory
Bookkeeping	Instant	Invoicing/accts payable
Purchasing	Instant	Research Materials/POs
Financial statements	4 weeks	Self explanatory
Grants Management	As needed	Self explanatory
Tax filings	PA/Fed dates	Certified Public Accountant firms being interviewed
KIZ filing[g]	biannual	PA tax refund program
Registration with NIH, DoD (SBIR compliance)[h]	2 weeks	Self explanatory
Deep IP Audit[i]	Instant	Is the technology protectable? Unique?
IP management[j]	Instant	From Invention disclosure, to Provisional, to full filings
Development & Product Plan[k]	4 weeks	Initial product plan should be a priority
Business Plan with market analysis[l]	4 weeks	Initial business plan with Newco formation
Begin SBIR application draft[h]	2 weeks	Self explanatory

[a]*Involvement of the Blumberg Institute faculty and staff inventors in decisions*: Management of appointments, development plans, and all initial planning should involve the participation and ideally the enthusiastic support of the faculty/staff inventors. Hopefully, the faculty/staff inventors are involved in the smooth operation of the Newco. However, if they wish to work in the Newco with compensated positions, arrangements must be made with the Blumberg Institute or HBF to avoid conflicts of effort and interest. Leaves of absence or sabbaticals, where appropriate, may be considered.

[b]*Technology Coordinator*: This is a technology transfer function, and the coordinator (or coordinator team) examines technologies generated by the Blumberg Institute faculty for commercialization potential by attending research seminars and reviewing annual reports. The technology

(continued)

Table 3.4 (continued)

coordinator or team also examines technologies at other research organizations for their commercialization potential and compatibility with programs at the Center. Technologies identified of being interest are then pursued. Working with the institution at which the inventions originate, the inventors or their staff members may be encouraged to come to the Center to develop their innovations commercially. Funding and other resources may be provided.

[c]*Overall management of Newcos*: During the initial period of Newco formation and operation, and usually through the first and second rounds of extramural investor funding, the Blumberg Institute will take an active role in coordinating Newco progress by convening regular meetings with Newco leadership. This will be supervised by the Blumberg Institute's Project and Portfolio Managers. The project manager ensures that each of the necessary activities to be performed on the punch list are carried out. The portfolio manager observes Newcos, in which the Blumberg Institute has made a cash investment. Both will report to the PABC CEO. If funding from the Blumberg Institute is provided, one of its representatives is expected to have an initial board seat, which could convert to an observer seat following new investor funding.

[d]*Newco incorporation and structure*: Working with the faculty/staff inventors, Newcos will be legally formed as an LLC or C-corporation with an ownership distribution according to the Blumberg Institute's company creation policy. The degree to which time and resources from the Blumberg Institute can be committed to the Newco is a function of the ownership model and agreements made with the institute.

[e]*Newco early management (initial leadership)*: Initial management may be full- or part-time and is often transitional. People in these positions may be compensated by (a) their regular compensation as the Blumberg Institute staff, (b) bonuses for this service, (c) equity for this service, or a combination of a, b, and c, (d) or strictly voluntary. The positions are likely to be transient to establish the Newco. As funds are raised, the transitional management may be invited to, and/or may wish to continue in their roles in the Newco. If the management personnel are the Blumberg Institute staff, they may take leave, resign, or make other plans to accommodate what could be a conflict of effort in their new position. Although inventors of the technology upon which the Newco is based may be vital in establishing the Newco, they are often not ideal leaders as the Newco matures. The Newco may be forced to make decisions that deprioritize the founding technology, and this could put the inventors in conflict; Board members, SAB, and other administrative staff initially serve part-time and for equity stakes, and are most likely drawn from the executive and expert network of individuals located onsite or affiliated with the PABC. A detailed spending and personnel plan should be drafted, as well as an extramural funding plan. This information will be considered in the business plan, but there will be a separate, parallel plan describing how the program will be funded, beginning with nondilutive sources, where appropriate (such as SBIR, STTR, state, and foundation support), and then Angel, VC, and other investors, culminating in various exits, including IPOs, acquisition, contracts, or sales.

[f]*Administration*: Many of these essential functions are to be carried out by the Blumberg Institute administration, and this will require appropriate conflict-of-interest and effort management, as well as fair compensation and consideration for the administrators and the institute.

[g,h]*KIZ and SAM Filing*: Filing as a KIZ company allows eligible Newcos to receive special concessions from the Commonwealth of PA, including tax credits and financial support. Filing with SAM.gov (System for Award Management) and the federal Grants.gov website may take months from filing to recognition of the Newco as valid by the government, which is necessary for application for federal funding, including SBIR, STTR, EDA and Defense Department grants. If appropriate, the Blumberg Institute Project Manager can work with the inventor/entrepreneur (or their designees) to help draft SBIR or STTR proposals. This presumes that a Newco has been created around a technology or one is anticipated.

[i,j]*Intellectual Property (IP) Audit and Protection Plan*: Once a technology is determined to be interesting and a good fit for development at the Center, it is referred to a patent attorney who is onsite at the PABC. They will perform an analysis to determine the patentability and patent integ-

(continued)

Table 3.4 (continued)

rity of the technology. This is followed by a patent protection plan. These plans need to be flexible and adapt to the changing environment in which the technology exists. New information generated and competing technologies that may emerge must be considered. There must be a balance between the costs of protection and the demand for the technology.

^k*Product Plan with Essential Feasibility Experiments*: The product plan will outline the critical work to be done to bring the technology from its present state into commercialization. This will include any laboratory work as well as advanced preclinical and clinical studies. Where outsourcing is anticipated, general plans will be included. A Target Product Profile should be drafted.

^l*Business and Market Plan*: This will be more of an Executive Summary, but should include the key elements present in a full business plan, such as startup company structure, assets, valuations, development and time charts, anticipated revenues and expenses, and even comments on reimbursement. This plan will be the basis for promotions for future SBIR and STTR grant proposals and investor pitches.

The first step in the technology development lifecycle of a startup at the PABC is the identification and evaluation of the technology. Technology evaluation is routinely and aggressively applied to the Blumberg Institute's faculty innovations. More recently, we have entered into agreements with research institutes and universities to allow us to examine their technology portfolios and determine if any of the technologies would be a good fit for development at the PABC. If the technology is determined to be best served by the creation of a Newco or its placement in an existing but early-stage Newco, a number of actions should be carried out. The punch list provided in Table 3.4 is intended to offer examples of these activities, noting that (a) the list is not all inclusive and (b) the PABC itself may not be the entity performing all of the needed functions. To a large extent, what is to be done is determined on a case-by-case basis, but in all circumstances, the establishment of a management team is considered to be an early and common action. The Blumberg Institute Project Manager and Portfolio Manager should coordinate their activities to successfully oversee the progress of each of the Newcos.

3.6 To Whom and For What Is the PABC Accountable?

The PABC is accountable to the Hepatitis B Foundation and the Commonwealth of Pennsylvania. The PABC was created by the HBF to be its home and to provide an environment in which the organization and its research interests could thrive. It is also intended to provide financial support to the HBF and its research organization, the Blumberg Institute. That mutualism is a key to its success, we believe. Importantly, the nonprofit PABC Board is appointed by the HBF. Therefore, there is a clear mechanism of accountability to the HBF. However, the mission and expectation of the community, which provides financial support to the PABC, is that there will be a beneficial impact on medical innovation, training, job creation and economic stimulation. It is important that those who invest in us are aware of our success. There are specific and general ways in which this can be done.

Relationships with community and legislative leaders: An intangible component of the PABC success, which is part cause and part effect, is its outstanding relationships with community leaders. Statewide, County and local representatives are all kept abreast of the PABC activities and contributions to the community. These elected representatives have become enthusiastic advocates for the Center, and many have actually had constructive ideas for the Center's development during roundtable discussions led by the PABC.

Accountability metrics: The PABC has commissioned regular assessments (every 3–5 years) of its impact on the region, the state, and the nation. Most specifically, our "Economic Impact" study conducted by independent economists provides objective analysis that can be shared with community leaders and others with an interest in the Center. Since the PABC receives funding from extramural and state sources, this is an effective way to provide accountability.

Economists generally categorize economic impact into various types, as defined by Weisbrod (1997) and summarized in a report from the University of Illinois (2018), paraphrased here:

Employment and Earnings Impact: The number of employees and their earnings, including non-cash and non-direct compensation.

Economic Output and Economic Activity: Gross receipts for goods and services generated by the enterprise.

Direct Impact: Spending on employment and goods and taxes paid by the entities within the enterprise in our community, the state, and the nation.

Indirect Impact: Spending made possible by the direct spending, which is a function of business-to-business transactions.

Induced Impact: Additional new spending and revenue made possible by the results of the direct and indirect impact, including effects on households.

Total Impact: Jobs, wages, and output in industries that result from direct and indirect employees, families, and effects. The total impact includes not only end-use consumer products but also any locally purchased intermediate products that were used in producing consumer goods and services.

Fiscal Impact of Operations: Includes the direct, indirect, induced, and total impacts of a new or existing company in terms of sales taxes, income taxes, realty transfer taxes, motor fuel taxes, gas taxes, vehicle licenses and fees, telecommunication and electricity excise taxes, property taxes, and other state and county revenues supported by the operations of the company. The operational phase impacts are generally considered long-term effects of the company.

Economic impact reports should be conducted by professionals specializing in these reports. Our report did not include all of the impact categories defined above. We were somewhat narrower, focusing on only (i) Direct, (ii) Indirect, and (iii) Induced Impact. We subcategorized the results further into impact upon the local region, entire state, and overall impact without regard to geography. The Association of University Research Parks (AURP) provides more information about the economic impact of their member organizations and can be a model and an additional resource.

3.7 Taking Equity in the Startup Companies at the PABC

In short, do it. However, justify doing so. In addition, note that there may be immediate tax filing implications for for-profit incubators taking equity. There is a point that the amount of for-profit and unrelated income can affect nonprofit status. However, that's for the tax experts. I cannot imagine any math where it doesn't make sense for the incubator to try to share in the upsides of the Newcos it nurtures.

According to a 2017 survey of 120 incubators in the U.S. from the InBIA (website), the average amount of equity in tenant companies taken by the incubator host is 6.8%. I think that incubator and certainly accelerator policy should be designed to take equity positions in the startups and Newcos they host, but they must be justified and in proportion to the value they provide. This is what I think is appropriate, but to be honest, for the first decade of operation of the PABC, we did not have a consistent policy about this.

When we opened the PABC in 2006, because of our partners' reluctance and my personal reticence, we never took equity in the Newcos, unless they were specifically created with our technologies. That meant we stayed on the sidelines and watched as many of the most exciting Newcos at the Center blossomed from nascent startups to billion-dollar valuations and half-billion dollar buyouts—without the PABC retaining any stake. Eventually, I sat down with the PABC tenant Leadership Council, a group of CEOs of the startups at the Center, most of whom are still my friends. I proposed that each of their companies *give* the PABC, or HBF (our parent nonprofit), 1% of their company. This was a somewhat arbitrary amount. The reasoning was that it's small enough to not have a disruptive impact on their "cap" tables but large enough to give our foundation mini-lottery tickets in the Newco without needing to justify or value each company. My thinking was they would be happy to do this, given the intangible contributions of the resources and environment we provide. Almost all of our tenants (stakeholders) have sung our praises, and we have had and continue to have excellent relations with each of them. Indeed, even without a policy or actual stake in the Newcos, the entrepreneurs have regularly made donations to our foundation, following their successful exits.

My proposal for a meager 1% stake, however, was not met with enthusiasm. There was considerable awkwardness. It was explained to me that the Newcos already had investors to whom they were accountable, and they could not, without a real value justification, give away any company shares (and I guess they weren't willing to do this as individuals). They did eventually agree to work with me in developing a value proposition, which we use today, and could be a model for other incubators.

Some incubators and accelerators make cash investments in their Newcos. For example, Y combinator provides $150,000–$500,000 in the form of a SAFE (simple agreement for future equity), which may end up being between 5 and 8% of the value of the Newco. The PABC may make cash investments using our Angel investment funding programs and expect equity stakes in exchange, but by policy we now have a justification-value proposition on the table for every Newco that enters the Center. The PABC was built with and is in part operated with donor and public money. This defrays the true cost of operation. We calculated what would be the cost per square foot of a rented space at the Center in the absence of this donor and public support and subtracted the amount actually paid. That difference, per year, is taken as a direct stake in the Newco, if a valuation is known, or as a convertible note in the Newco, delaying valuation until specified funding rounds.

3.8 Graduate Space and Replicating the Model

The PABC has been used as a model for incubator design, operation, and corridor development influence in part of the country. For example, working with municipal leadership from the township, county, and state, the PABC is generating plans to develop the immediate area around the Center into a technology corridor in which there are a cluster of biotech and high-tech companies within short distances of each other. This is intentionally ambitiously and analogous to Cambridge's Kendell Square and Philadelphia's University Sciences City Center. It has been, of course, scaled and adapted to the PABC culture and resources. These plans imagine a knowledge community of thousands of skilled professionals working in dozens of companies that grow from our incubator or are now attracted to the area because of the emergence of a talent pool and infrastructure. We specifically anticipate working with real estate developers to build "graduate space" immediately adjacent to the PABC for companies that outgrow our available space. This would be the first step in corridor development, followed by further development of nearby properties. We are also actively working with public and private agencies to build and manage business incubators in other parts of the U.S. These activities will be briefly summarized below.

Graduate space for PABC companies and a knowledge ecosystem in Bucks County, PA: Startups that become successful and grow will often suddenly need more space. For many successful startups there comes a time when it is simply not practical to remain within our campus. Often, program success is accompanied by major investments and there is rapid, explosive growth. We usually scramble into action and do whatever we can to find space on our campus. If this fails, the PABC works with public and private organizations to find what we have been calling "graduate space." These are usually commercial facilities that can accommodate research-based companies needing 10,000–50,000 ft^2 of lab, office, and production space. Historically, we scour the county looking for the closest available facilities. These growing Newcos can be the building blocks of a Bucks County tech corridor that radiates from the PABC. The growth of these Newcos provides justification for the construction of new facilities adjacent to us. The construction of these units, which are currently being planned to be within walking distance of the core campus, is part of a strategy to build an "innovation corridor." The reality and success of this innovation corridor depends on the willingness of the growing Newcos to remain nearby, even after their explosive growth. This growth becomes a catalyst for more growth. We become a magnet to attract biotech and related companies from all over the country.

There is a balance between hanging on to Newcos for the good of the PABC and encouraging them to leave for their own good. Ideally, the Newcos in the PABC will grow and the need for more space on PABC's 14.5 acres will become inevitable. By custom, the PABC prefers to retain companies that are successfully progressing as long as possible and practical. Companies that have matured can provide resources, mentoring, deals, and modeling to the other startup entrepreneurs just arriving. It is a terrifically effective ecosystem.

However, creating options for them to locate near the PABC is a way in which there can be benefits for all parties involved beyond the time that their accommodation on the campus is possible. We have found that many of these rapidly growing companies wish to keep staff and even labs at the PABC after they expand their operations to sites off the PABC campus. This makes their growth nearby all the more justified. These companies find that keeping a group at the PABC allows for continuity of operation, continued access to scholarly activities, and even infrastructural resources, although equipment access usually becomes much less necessary for growing companies. Frankly, it appears that the founders enjoy keeping a footprint at the PABC. In addition, this is great for the PABC, too. Companies that retain a presence have the virtue of maintaining relationships well until the time that the expanding companies have grown to multi-hundred employee operations.

3.9 Taking the PABC Model On the Road: Replicating the Model

Reproducing the Model
I always thought we had a remarkable strategy for success for the PABC. Less certain was the extent to which this strategy would be reproducible. I am proud of what was done here in Doylestown PA, but I had never been much of a student of biotech centers. Nevertheless, we were frequently approached by real estate developers and representatives from governmental organizations from around the country about helping or becoming involved in creating biotech business incubators. I usually demurred. Lou Kassa, our EVP and COO, at the time, was much more confident about what we had to offer. Today, as CEO he has a national vision of what can be done and the know-how to get this accomplished. He discussed these ideas with a number of different real estate developers and ultimately consummated an

agreement with Jerry Sweeny, CEO of The Brandywine Realty Trust to create and manage a life sciences incubator – called B Labs – in the 35-story skyscraper Cira Center in Philadelphia.

The Brandywine Realty Trust is a multibillion dollar publicly traded real estate development company that is one of the largest developers of office and commercial space in the Philadelphia region. It is expanding around the country. Until meeting us, it is my understanding that they had done little in the area of life science and biotechnology incubators, yet they had significant physical footprints near the research hubs of the University of Pennsylvania and Drexel University in West Philadelphia. That area is a hotbed of life sciences and biotechnology, and Mr. Sweeny was clearly predisposed to expanding into lab space. Life science business incubators, however, are an entirely different kind of operation than commercial lab space.

With an introduction from Tim Kelly, who had been the builder of our new 38,000 ft^2 building in Doylestown, Lou took the lead in conceiving and bringing to reality a business (life sciences) incubator on the first three floors (above the main lobby) of the Cira Center. This is a partnership between the PABC, the Blumberg Institute and the Brandywine Realty Trust. We introduced them to the world of lab-fitted space incubators. Our partnership with them is to help conceive and manage incubators, and Brandywine is now building hundreds of thousands of square feet of commercial lab space. This is quite an endorsement of our business strategy and shows their confidence in lab space development. The Brandywine incubators, in which we are partners, are called "B Labs," in recognition of all of the B's involved (Brandywine, PABC, Blumberg Institute, and HBF). The B Labs in Philadelphia is expected to be the first in a series around the country.

The idea, of which Lou is the champion, is to create incubators for startups stemming from nearby academic institutions with a culture that nurtures co-collaboration, as has become the term of art. We do this by planting nonprofit researchers in the life sciences incubator, and this model, discussed further in the next section, is the basis of the PABC that has been nicely replicated and, frankly, enhanced at B Labs.

The B Labs were quite a risk for Brandywine, but even at opening, which was in the middle of the COVID-19 pandemic, the incubator was 70% occupied. Within 6 months it was 100% occupied with a waiting list. Mr. Sweeny is a sought-after business speaker in the Philadelphia area, and Lou and I attended one of his presentations. Turning his attention to Lou and me in the audience, I heard Mr. Sweeny say how appreciative he was that he became involved with biotech and labs because, during the pandemic, when office occupancy in his properties declined significantly and lease renewals were drying up (my words), his lab spaces never dipped. Mr. Sweeny was generous in recognizing us, but I will be the first to say that I have never been confident about replicating the model, and the energy to do this has all been Lou.

With the PABC a success in two of the most different environments that can be imagined—an old warehouse in Doylestown and a modern skyscraper in Philadelphia — the question is, where next? There has been considerable interest in replication of the PABC model beyond the B Labs in Philadelphia and taking it to other places in the U.S. There are strategic and financial incentives to the PABC to model replication, especially where the PABC is a partner in the expansion. Clearly, life science business incubators can play a useful role in nurturing and enhancing the likelihood of the success of Newcos, especially when these innovations come from academia. Making a nonprofit research organization a core occupant of the business incubator and trusting it to be the manager also appears to confer a unique and productive environment that is particularly effective. Can this all be reproduced similarly to other franchises? That does seem possible but the success of this approach and culture depends upon the nonprofit organization, geography, demography, and the innovators and their technology.

References

Albort-Morant G, Oghazi P (2016) How useful are incubators for new entrepreneurs? J Bus Res 69(6):2125–2129

Azoulay P, Jones BF, Kim JD, Miranda J (2020) Age and high-growth entrepreneurship. Am Econ Rev Insights 2(1):65–82

Brieger SA, Bäro A, Criaco G, Terjesen SA (2021) Entrepreneurs' age, institutions, and social value creation goals: a multicountry study. Small Bus Econ 57:425–453

Clark C, Johnson H, Mas P (2018) Santa Rosa County business incubator (report)

de Bem Machado A, Catapan AH, Sousa MJ (2020) Management models for business incubators: a systematic review. Int J Technol Diffusion (IJTD) 11(2):33–44

Fern D, Arruti A, Markuerkiaga L, Saenz N (2018) The entrepreneurial university: a selection of good practices. J Entrepreneurship Educ 21:1–17

Fernández ND, Arruti A, Markuerkiaga L, Nerea S (2018) The entrepreneurial university: a selection of good practices. J Entrepreneurship Educ 21:1–17

Hackett S, Dilts DM (2008) Inside the black box of business incubation: study B – scale assessment, model refinement, and incubation outcomes. J Technol Transfer 33(5):439–471

Mubarak Al-Mubaraki H, Busler M (2010) Business incubators models of the USA and UK: a SWOT analysis. World J Entrepreneurship Manag Sustain Dev 6(4):335–354

Weisbrod BA (1997) The future of the nonprofit sector: its entwining with private enterprise and government. J Policy Anal Manag 16(4):541–555

Afterword

There are certainly many ways to nurture innovation. One thing is clear, though. Some people, some projects and some ideas may need little help beyond the genius of the inventor. In life sciences, someone who can have an idea and turn it into a real applied product is uncommon. Life sciences incubators, therefore, play a special role in bringing innovations to the public good. At the Pennsylvania Biotechnology Center (PABC), we leveraged the interest, usefulness and economics of business incubators to the good of a mission-driven nonprofit organization. The incubator we created is an enormous success, and it converts that success into contributions to the mission-oriented Hepatitis B Foundation to "codify" a new means of businesses linked to philanthropy. Thus far, so good, but future success is never assured.

And something else has been on my mind. We started the Hepatitis B Foundation (HBF) and the Baruch S. Blumberg Institute to find a cure and improve the lives of people affected by hepatitis B worldwide. There is no question. The HBF has been a valuable, trusted portal of information about hepatitis B and liver cancer for tens of thousands of people every year. The HBF is generally the only voice in Washington, D.C., advocating for those affected by this chronic liver disease.

But what about a cure for HBV? I started the Blumberg Institute and the PABC to find the cure. Perhaps Dr. Baruch Blumberg was right after all. To hit the mark you should not aim directly at the target. So, although there is no complete HBV cure as of yet, there is now a wave of new drugs on the horizon being tested for treating chronic hepatitis B. Some are very promising. And we can take pride in knowing that nearly 50% of those new investigational drugs have critical connections to the PABC. It's true. Almost 50%, ranging from the trendy siRNA inhibitors to HBV surface and capsid inhibitors. Scientists from the Blumberg Institute were directly responsible for some and companies at the PABC were critical for others. In one way or another, the new HBV drugs or the inventions leading to them had a foot here. Creating a biotech center may not be the most efficient way to cure a disease that is not on the public's radar screen, but it can definitely work and produce a great deal of good from the ground up for other diseases as well.

Appendices

Appendix A

The Formula for PABC Success as a List

PABC's key components for our success:
Research non profits at the PABC and at the helm
Research non profits at the PABC, generating ideas and innovations and new companies
Selection of companies that are a good fit
Providing a stimulating knowledge community
Providing common resources
Offering interventions to help with funding and networking
Excellent relationships with community and legislative leaders
Accountability, commissioning regular impact studies

Appendix B

Timeline for PABC

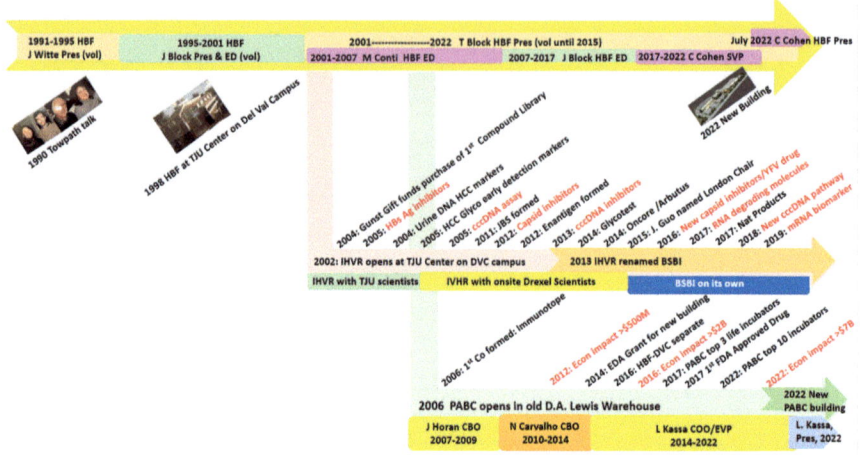

Index

A
Accelerators, 47, 51, 67, 71–73, 75, 93, 101, 102

B
Biotechnology, 2, 7, 8, 18–31, 33, 39, 51–79, 83–105

E
Entrepreneurship, 9, 10, 19, 27, 51–79, 83
Equity, 42, 72, 75, 76, 95, 98, 101–102

F
Failure, 22, 67, 78, 79

H
Hepatitis B, 2–9, 18, 20, 21, 24, 25, 33, 37, 40, 84, 88, 99
HVAC, 86, 87

I
Incubators, 1–49, 51–79, 83–89, 91, 93–96, 101–105
Infrastructures, 51, 71, 72, 85–87, 96, 102
Initial public offerings (IPOs), 42, 61, 77, 83, 94, 98
Innovations, 10, 44, 51–62, 68, 69, 71, 75–77, 79, 91, 95, 96, 98, 99, 103, 105

L
Leases, 13–15, 84, 104

M
Management, 73, 75, 79, 87–92, 94, 96–99

N
Nonprofits, 3, 4, 6, 7, 9–11, 13, 15–19, 24, 28, 32–34, 40–42, 44, 67, 68, 71, 75–76, 83, 84, 88, 89, 91, 96, 99, 101, 104, 105, 109

P
Pandemic, 45–49, 52, 53, 78, 104

R
Recession, 30, 31, 35

S
Small businesses, 2, 55–62, 76, 78
Start-up, 65, 66
Successes, 18, 30, 39, 42, 67, 73–79, 83, 84, 89, 91, 93, 99, 100, 102–105, 109–110

U
Universities, 1, 3, 8–13, 16, 21, 23–29, 33–36, 38, 46, 52, 56, 68–70, 74, 75, 83, 92, 94–96, 99–102, 104

© The Editor(s) (if applicable) and The Author(s), under exclusive license to
Springer Nature Switzerland AG 2024
T. M. Block, *Curing Disease from the Ground Up*,
https://doi.org/10.1007/978-3-031-56148-1

MIX
Papier aus verantwortungsvollen Quellen
Paper from responsible sources
FSC® C105338

If you have any concerns about our products,
you can contact us on
ProductSafety@springernature.com

In case Publisher is established outside the EU,
the EU authorized representative is:
**Springer Nature Customer Service Center GmbH
Europaplatz 3, 69115 Heidelberg, Germany**

Printed by Libri Plureos GmbH
in Hamburg, Germany